REAL-WORLD PYTHON

REAL-WORLD PYTHON

PYTHON

A Hacker's Guide to Solving Problems with Code

by Lee Vaughan

no starch press

San Francisco

Printed in USA

First printing

24 23 22 21 20   1 2 3 4 5 6 7 8 9

ISBN-13: 978-1-7185-0062-4 (print)
ISBN-13: 978-1-7185-0063-1 (ebook)

Publisher: William Pollock
Executive Editor: Barbara Yien
Production Editor: Kassie Andreadis
Developmental Editor: Frances Saux
Project Editor: Dapinder Dosanjh
Cover Illustrator: Rob Gale
Interior Design: Octopod Studios
Technical Reviewers: Chris Kren and Eric Mortenson
Copyeditor: Kim Wimpsett
Compositor: Shawn Morningstar
Proofreader: Paula L. Fleming
Indexer: Beth Nauman-Montana

The following images are reproduced with permission: Figure 3-3 from istockphoto.com; Figure 5-1 courtesy of Lowell Observatory Archives; Figures 5-2, 6-2, 7-6, 7-7, 8-18, and 11-2 courtesy of Wikimedia Commons; Figures 7-2, 7-9, 7-17, 8-20, and 11-1 courtesy of NASA; Figure 8-1 photo by Evan Clark; Figure 8-4 photo by author; Figure 9-5 from pixelsquid.com; Figure 11-9 photo by Hannah Vaughan

For information on distribution, translations, or bulk sales, please contact No Starch Press, Inc. directly:
No Starch Press, Inc.
245 8th Street, San Francisco, CA 94103
phone: 1-415-863-9900; info@nostarch.com
www.nostarch.com

Library of Congress Cataloging-in-Publication Data

Names: Vaughan, Lee, author.
Title: Real-world python: a hacker's guide to solving problems with code / Lee Vaughan.
Description: San Francisco, CA : No Starch Press, Inc., [2020] | Includes
    index.
Identifiers: LCCN 2020022671 (print) | LCCN 2020022672 (ebook) | ISBN
    9781718500624 (paperback) | ISBN 1718500629 (paperback) | ISBN
    9781718500631 (ebook)
Subjects: LCSH: Python (Computer program language)
Classification: LCC QA76.73.P98 V383 2020 (print) | LCC QA76.73.P98
    (ebook) | DDC 005.1/33--dc23
LC record available at https://lccn.loc.gov/2020022671
LC ebook record available at https://lccn.loc.gov/2020022672

For my uncle, Kenneth P. Vaughan.
He brightened every room he entered.

About the Author

Lee Vaughan is a programmer, pop culture enthusiast, educator, and author of *Impractical Python Projects* (No Starch Press, 2018). As an executive-level scientist at ExxonMobil, he constructed and reviewed computer models, developed and tested software, and trained geoscientists and engineers. He wrote both *Impractical Python Projects* and *Real-World Python* to help self-learners hone their Python skills and have fun doing it!

About the Technical Reviewers

Chris Kren graduated from the University of South Alabama with an M.S. in Information Systems. He currently works in the field of cybersecurity and often uses Python for reporting, data analysis, and automation.

Eric Mortenson has a PhD in mathematics from the University of Wisconsin at Madison. He has held research and teaching positions at The Pennsylvania State University, The University of Queensland, and the Max Planck Institute for Mathematics. He is an associate professor in mathematics at St. Petersburg State University.

BRIEF CONTENTS

CONTENTS IN DETAIL

9
IDENTIFYING FRIEND OR FOE   203

10
RESTRICTING ACCESS WITH FACE RECOGNITION   225

11
CREATING AN INTERACTIVE ZOMBIE ESCAPE MAP   245

12
ARE WE LIVING IN A COMPUTER SIMULATION?   269

APPENDIX
PRACTICE PROJECT SOLUTIONS   283

ACKNOWLEDGMENTS

Despite operating during a global pandemic, the team at No Starch Press delivered another excellent effort at book making. They are professionals without peer, and this book would not exist without them. They have my deepest gratitude and respect.

Thanks also to Chris Kren and Eric Evenchick for their code reviews, Joseph B. Paul and Sarah and Lora Vaughan for their cosplay enthusiasm, and Hannah Vaughan for supplying useful photographs.

Special thanks to Eric T. Mortenson for his meticulous technical reviews and many helpful suggestions and additions. Eric proposed the chapter on Bayes' Rule and supplied numerous practice and challenge projects including applying Monte Carlo simulation to Bayes, summarizing a novel by chapter, modeling interactions between the moon and Apollo 8, viewing Mars in 3D, calculating the light curve for an exoplanet with an orbiting moon, and more. This book is immensely better for his efforts.

Finally, thanks to all the contributors to *stackoverflow.com*. One of the best things about Python is its extensive and inclusive user community. No matter what question you may have, someone can answer it; no matter what strange thing you want to do, someone has probably done it before, and you can find them on Stack Overflow.

INTRODUCTION

If you've learned the basics of coding in Python, you're ready to write complete programs that take on real-world tasks.

In *Real-World Python*, you'll write programs to win the moon race with Apollo 8, help Clyde Tombaugh discover Pluto, select landing sites for a Mars rover, locate exoplanets, send super-secret messages to your friends, battle monstrous mutants, save shipwrecked sailors, escape the walking dead, and more, all using the Python programming language. In the process, you'll apply powerful computer vision, natural language processing, and scientific modules, such as OpenCV, NLTK, NumPy, pandas, and matplotlib, as well as a host of other packages designed to make your computing life easier.

Who Should Read This Book?

You can think of this as a sophomore Python book. It isn't a tutorial on programming basics but rather a way for you to continue training using a project-based approach. This way, you won't have to waste your money or shelf space rehashing concepts you've already learned. I'll still explain every step of the projects, and you'll receive detailed instructions about using the libraries and modules, including how to install them.

These projects will appeal to anyone who wants to use programming to conduct experiments, test theories, simulate nature, or just have fun. As you work through them, you'll increase your knowledge of Python libraries and modules and learn handy shortcuts, useful functions, and helpful techniques. Rather than focus on isolated modular code snippets, these projects teach you how to build complete, working programs involving real-world applications, datasets, and issues.

Why Python?

Python is a high-level, interpretive, general-purpose programming language. It's free, highly interactive, and portable across all major platforms and microcontrollers such as the Raspberry Pi. Python supports both functional and object-oriented programming and can interact with code written in many other programming languages, such as C++.

Because Python is accessible to beginners and useful to experts, it has penetrated schools, universities, large corporations, financial institutions, and most, if not all, fields of science. As a result, it's now the most popular language for machine learning, data science, and artificial intelligence applications.

What's in This Book?

The following is an overview of the chapters in this book. You don't have to work through them sequentially, but I'll explain new modules and techniques more thoroughly when they're first introduced.

Chapter 1: Saving Shipwrecked Sailors with Bayes' Rule Use Bayesian probability to efficiently direct Coast Guard search and rescue efforts off Cape Python. Gain experience with OpenCV, NumPy, and the itertools module.

Chapter 2: Attributing Authorship with Stylometry Use natural language processing to determine whether Sir Arthur Conan Doyle or H. G. Wells wrote the novel *The Lost World*. Gain experience with NLTK, matplotlib, and stylometric techniques such as stop words, parts of speech, lexical richness, and Jaccard similarity.

Chapter 3: Summarizing Speeches with Natural Language Processing Scrape famous speeches off the internet and automatically produce a summary of the salient points. Then turn the text of a novel into a cool display for advertising or promotional material. Gain experience with `BeautifulSoup`, `Requests`, `regex`, NLTK, `Collections`, `wordcloud`, and `matplotlib`.

Chapter 4: Sending Super-Secret Messages with a Book Cipher   Share unbreakable ciphers with your friends by digitally reproducing the one-time pad approach used in Ken Follet's best-selling spy novel, *The Key to Rebecca*. Gain experience with the `Collections` module.

Chapter 5: Finding Pluto   Reproduce the blink comparator device used by Clyde Tombaugh to discover Pluto in 1930. Then use modern computer vision techniques to automatically find and track subtle transients, such as comets and asteroids, moving against a starfield. Gain experience with OpenCV and `NumPy`.

Chapter 6: Winning the Moon Race with Apollo 8   Take the gamble and help America win the moon race with Apollo 8. Plot and execute the clever free return flight path that convinced NASA to go to the moon a year early and effectively killed the Soviet space program. Gain experience using the `turtle` module.

Chapter 7: Selecting Martian Landing Sites   Scope out potential landing sites for a Mars lander based on realistic mission objectives. Display the candidate sites on a Mars map, along with a summary of site statistics. Gain experience with OpenCV, the Python Imaging Library, `NumPy`, and `tkinter`.

Chapter 8: Detecting Distant Exoplanets   Simulate an exoplanet's passing before its sun, plot the resulting changes in relative brightness, and estimate the diameter of the planet. Finish by simulating the direct observation of an exoplanet by the new James Webb Space Telescope, including estimating the length of the planet's day. Use OpenCV, `NumPy`, and `matplotlib`.

Chapter 9: Identifying Friend or Foe   Program a robot sentry gun to visually distinguish between Space Force Marines and evil mutants. Gain experience with OpenCV, `NumPy`, `playsound`, `pyttsxw`, and `datetime`.

Chapter 10: Restricting Access with Face Recognition   Restrict access to a secure lab using face recognition. Use OpenCV, `NumPy`, `playsound`, `pyttsxw`, and `datetime`.

Chapter 11: Creating an Interactive Zombie Escape Map   Build a population density map to help the survivors in the TV show *The Walking Dead* escape Atlanta for the safety of the American West. Gain experience with `pandas`, `bokeh`, `holoviews`, and `webbrowser`.

Chapter 12: Are We Living in a Computer Simulation?   Identify a way for simulated beings—perhaps us—to find evidence that they're living in a computer simulation. Use `turtle`, `statistics`, and `perf_counter`.

Each chapter ends with at least one practice or challenge project. You can find solutions to the practice projects in the appendix or online. These aren't the only solutions, or necessarily the best ones; you may come up with better ones on your own.

When it comes to the challenge projects, however, you're on your own. It's sink or swim, which is a great way to learn! My hope is that this book motivates you to create new projects, so think of the challenge projects as seeds for the fertile ground of your own imagination.

You can download all of the book's code, including solutions to the practice projects, from the book's website at *https://nostarch.com/real-world-python/*. You'll also find the errata sheet there, along with any other updates.

It's almost impossible to write a book like this without some initial errors. If you see a problem, please pass it on to the publisher at *errata@nostarch.com*. We'll add any necessary corrections to the errata and include the fix in future printings of the book, and you will gain eternal glory.

Python Version, Platform, and IDE

I built all the projects in this book with Python v3.7.2 in a Microsoft Windows 10 environment. If you're using a different operating system, no problem: I suggest compatible modules for other platforms, where appropriate.

The code examples in this book are from either the Python IDLE text editor or the interactive shell. IDLE stands for *integrated development and learning environment*. It's an *integrated development environment (IDE)* with an *L* added so that the acronym references Eric Idle of *Monty Python* fame. The interactive shell, also called the *interpreter*, is a window that lets you immediately execute commands and test code without needing to create a file.

IDLE has numerous drawbacks, such as the lack of a line-number column, but it's free and bundled with Python, so everyone has access to it. You're welcome to use whichever IDE you want. Popular choices include Visual Studio Code, Atom, Geany (pronounced "genie"), PyCharm, and Sublime Text. These work with a wide range of operating systems, including Linux, macOS, and Windows. Another IDE, PyScripter, works only with Windows. For an extensive listing of available Python editors and compatible platforms, visit *https://wiki.python.org/moin/PythonEditors/*.

Installing Python

You can choose to install Python directly on your machine or through a distribution. To install directly, find the installation instructions for your operating system at *https://www.python.org/downloads/*. Linux and macOS machines usually come with Python preinstalled, but you may want to upgrade this installation. With each new Python release, some features are added and some are deprecated, so I recommend upgrading if your version predates Python v3.6.

The download button on the Python site (Figure 1) may install 32-bit Python by default.

Figure 1: Downloads page for Python.org, with the "easy button" for the Windows platform

If you want the 64-bit version, scroll down to the listing of specific releases (Figure 2) and click the link with the same version number.

Looking for a specific release?

Python releases by version number:

Release version	Release date		Click for more
Python 3.7.7	March 10, 2020	⬇ Download	Release Notes
Python 3.8.2	Feb. 24, 2020	⬇ Download	Release Notes
Python 3.8.1	Dec. 18, 2019	⬇ Download	Release Notes
Python 3.7.6	Dec. 18, 2019	⬇ Download	Release Notes
Python 3.6.10	Dec. 18, 2019	⬇ Download	Release Notes
Python 3.8.0	Nov. 2, 2019	⬇ Download	Release Notes
Python 3.5.5	Oct. 29, 2019	⬇ Download	Release Notes

Figure 2: Listing of specific releases from the Python.org downloads page

Clicking the specific release will take you to the screen shown in Figure 3. From here, click the 64-bit executable installer, which will launch an installation wizard. Follow the wizard directions and take the default suggestions.

Files

Version	Operating System	Description	MD5 Sum	File Size	GPG
Gzipped source tarball	Source release		3ee10f25e3d1b14215d56c3882466fcf	22973527	SIG
XZ compressed source tarball	Source release		93df27aec0cd18d6d42173e601ffb0fd	17108364	SIG
macOS 64-bit/32-bit installer	Mac OS X	for Mac OS X 10.6 and later	5a085f27f15e0d600de28d6232c0b6954	34479513	SIG
macOS 64-bit installer	Mac OS X	for OS X 10.9 and later	4ca0e30f46be690bfe80111daee9509a	27039889	SIG
Windows help file	Windows		7740b11d249bca16364f4a45b40c9676	8090273	SIG
Windows x86-64 embeddable zip file	Windows	for AMD64/EM64T/x64	854ac011903b4c798379a3baa3a040ec	7018568	SIG
Windows x86-64 executable installer	Windows	for AMD64/EM64T/x64	a2b79563476e9ee47f11899a63349383	26190920	SIG
Windows x86-64 web-based installer	Windows	for AMD64/EM64T/x64	047d19d2560c963b8251a9b2e52395ef	1362888	SIG
Windows x86 embeddable zip file	Windows		70df01ef9d6c1b7042aabb5a3c1e2fbd5	6526496	SIG
Windows x86 executable installer	Windows		ebf1644cdc1eeeebacc92efa849ck01	25424128	SIG
Windows x86 web-based installer	Windows		d3944e218a45d982f0abcd93b151271a	1324632	SIG

Figure 3: File listing for Python 3.8.2 version on Python.org

Some of the projects in this book call for nonstandard packages that you'll need to install individually. This isn't difficult, but you can make things easier by installing a Python distribution that efficiently loads and manages hundreds of Python packages. Think of this as one-stop shopping. The package managers in these distributions will automatically find and download the latest version of a package, including all of its dependencies.

Anaconda is a popular free distribution of Python provided by Continuum Analytics. You can download it from *https://www.anaconda.com/.* Another is Enthought Canopy, though only the basic version is free. You can find it at *https://www.enthought.com/product/canopy/.* Whether you install Python and its packages individually or through a distribution, you should encounter no problems working through the projects in the book.

Running Python

After installation, Python should show up in your operating system's list of applications. When you launch it, the shell window should appear (shown in the background of Figure 4). You can use this interactive environment to run and test code snippets. But to write larger programs, you'll use a text editor, which lets you save your code, as shown in Figure 4 (foreground).

Figure 4: The native Python shell window (background) and text editor (foreground)

To create a new file in the IDLE text editor, click **File ▸ New File**. To open an existing file, click **File ▸ Open** or **File ▸ Recent Files**. From here, you can run your code by clicking **Run ▸ Run Module** or by pressing F5 after clicking in the editor window. Note that your environment may look different from Figure 4 if you chose to use a package manager like Anaconda or an IDE like PyCharm.

You can also start a Python program by typing the program name in PowerShell or Terminal. You'll need to be in the directory where your Python program is located. For example, if you didn't launch the Windows PowerShell from the proper directory, you'll need to change the directory path using the cd command (Figure 5).

Figure 5: Changing directories and running a Python program in the Windows PowerShell

To learn more, see *https://pythonbasics.org/execute-python-scripts/*.

Using a Virtual Environment

Finally, you may want to install the dependencies for each chapter in a separate virtual environment. In Python, a *virtual environment* is a self-contained directory tree that includes a Python installation and a number of additional packages. They're useful when you have multiple versions of Python installed, as some packages may work with one version but break with others. Additionally, it's possible to have projects that need different versions of the same package. Keeping these installations separate prevents compatibility issues.

The projects in this book don't require the use of virtual environments, and if you follow my instructions, you'll install the required packages system-wide. However, if you do need to isolate the packages from your operating system, consider installing a different virtual environment for each chapter of the book (see *https://docs.python.org/3.8/library/venv.html#module-venv* and *https://docs.python.org/3/tutorial/venv.html*).

Onward!

Many of the projects in this book rely on statistical and scientific concepts that are hundreds of years old but impractical to apply by hand. But with the introduction of the personal computer in 1975, our ability to store, process, and share information has increased by many orders of magnitude.

In the 200,000-year history of modern humans, only those of us living in the last 45 years have had the privilege of using this magical device and realizing dreams long out of reach. To quote Shakespeare, "We few. We happy few."

Let's make the most of the opportunity. In the pages that follow, you'll easily accomplish tasks that frustrated past geniuses. You'll scratch the surface of some of the amazing feats we've recently achieved. And you might even start to imagine discoveries yet to come.

1

SAVING SHIPWRECKED SAILORS WITH BAYES' RULE

Sometime around 1740, an English Presbyterian minister named Thomas Bayes decided to mathematically prove the existence of God. His ingenious solution, now known as *Bayes' rule*, would become one of the most successful statistical concepts of all time. But for 200 years it languished, largely ignored, because its tedious mathematics were impractical to do by hand. It took the invention of the modern computer for Bayes' rule to reach its full potential. Now, thanks to our fast processors, it forms a key component of data science and machine learning.

Because Bayes' rule shows us the mathematically correct way to incorporate new data and recalculate probability estimates, it penetrates almost all human endeavors, from cracking codes to picking presidential winners to demonstrating that high cholesterol causes heart attacks. A list of applications of Bayes' rule could easily fill this chapter. But since nothing is more important than saving lives, we'll focus on the use of Bayes' rule to help save sailors lost at sea.

In this chapter, you'll create a simulation game for a Coast Guard search and rescue effort. Players will use Bayes' rule to guide their decisions so they can locate the sailor as quickly as possible. In the process, you'll start working with popular computer vision and data science tools like Open Source Computer Vision Library (OpenCV) and NumPy.

Bayes' Rule

Bayes' rule helps investigators determine the probability that something is true given new evidence. As the great French mathematician Laplace put it, "The probability of a cause—given an event—is proportional to the probability of the event—given its cause." The basic formula is

$$P(A/B) = \frac{P(B/A)\ P(A)}{P(B)}$$

where A is a hypothesis and B is data. $P(A/B)$ means the probability of A given B. $P(B/A)$ means the probability of B given A. For example, assume we know that a certain test for a certain cancer is not always accurate and can give false positives, indicating that you have cancer when you don't. The Bayes expression would be

$$\left(\begin{array}{c}\text{Probability of cancer}\\\text{given a positive test}\end{array}\right) = \left(\begin{array}{c}\text{Probability of a positive test}\\\text{among cancer patients}\end{array}\right) \times \frac{\left(\begin{array}{c}\text{Probability of}\\\text{having cancer}\end{array}\right)}{\left(\begin{array}{c}\text{Probability of}\\\text{a positive test}\end{array}\right)}$$

The initial probabilities would be based on clinical studies. For example, 800 out of 1,000 people who have cancer may receive a positive test result, and 100 out of 1,000 may be misdiagnosed. Based on disease rates, the overall chance of a given person having cancer may only be 50 out of 10,000. So, if the overall probability of having cancer is low and the overall probability of getting a positive test result is relatively high, the probability of having cancer given a positive test goes down. If studies have recorded the frequency of inaccurate test results, Bayes' rule can correct for measurement errors!

Now that you've seen an example application, look at Figure 1-1, which shows the names of the various terms in Bayes' rule, along with how they relate to the cancer example.

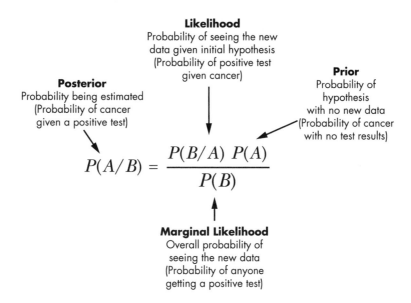

Figure 1-1: Bayes' rule with terms defined and related to the cancer test example

To illustrate further, let's consider a woman who has lost her reading glasses in her house. The last time she remembers wearing them, she was in her study. She goes there and looks around. She doesn't see her glasses, but she does see a teacup and remembers that she went to the kitchen. At this point, she must make a choice: search the study more thoroughly or leave and check the kitchen. She decides to go to the kitchen. She has unknowingly made a Bayesian decision.

She went to the study first because she felt it offered the highest probability for success. In Bayesian terms, this initial probability of finding the glasses in the study is called the *prior*. After a cursory search, she changed her decision based on two new bits of information: she did not easily find the glasses, and she saw the teacup. This represents a *Bayesian update*, in which a new posterior estimate ($P(A/B)$ in Figure 1-1) is calculated as more evidence becomes available.

Let's imagine that the woman decided to use Bayes' rule for her search. She would assign actual probabilities both to the likelihood of the glasses being in either the study or the kitchen and to the effectiveness of her searches in the two rooms. Rather than intuitive hunches, her decisions are now grounded in mathematics that can be continuously updated if future searches fail.

Figure 1-2 illustrates the woman's search for her glasses with these probabilities assigned.

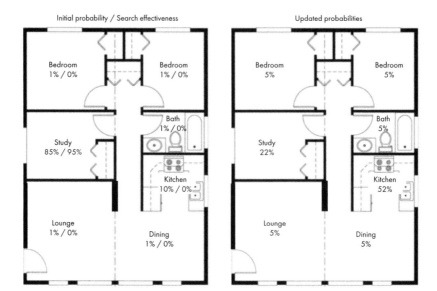

Figure 1-2: Initial probabilities for the location of the glasses and search effectiveness (left) versus updated target probabilities for the glasses (right)

The left diagram represents the initial situation; the right diagram is updated with Bayes' rule. Initially, let's say there was an 85 percent chance of finding the glasses in the study and a 10 percent chance that the glasses are in the kitchen. Other possible rooms are given 1 percent because Bayes' rule can't update a target probability of zero (plus there's always a small chance the woman left them in one of the other rooms).

Each number after a slash in the left diagram represents the *search effectiveness probability (SEP)*. The SEP is an estimate of how effectively you've searched an area. Because the woman has searched only in the study at this point, this value is zero for all other rooms. After the Bayesian update (the discovery of the teacup), she can recalculate the probabilities based on the search results, shown on the right. The kitchen is now the most likely place to look, but the probability for the other rooms increases as well.

Human intuition tells us that if something isn't where we think it is, the odds that it is someplace else go up. Bayes' rule takes this into account, and thus the probability that the glasses are in other rooms increases. But this can happen only if there was a chance of them being in the other room in the first place.

The formula used for calculating the probability that the glasses are in a given room, given the search effectiveness, is

$$P(G/E) = \frac{P(E/G)P_{\mathrm{prior}}(G)}{\Sigma P(E/G')P_{\mathrm{prior}}(G')}$$

where G is the probability that the glasses are in a room, E is the search effectiveness, and P_{prior} is the prior, or initial, probability estimate before receiving the new evidence.

You can obtain the updated possibility that the glasses are in the study by inserting the target and search effectiveness probabilities into the equation as follows:

$$\frac{\left(0.85 \times (1 - 0.95)\right)}{\left(0.85 \times (1 - 0.95) + 0.1 \times (1 - 0) + 0.01 \times (1 - 0) + 0.01 \times (1 - 0) + 0.01 \times (1 - 0) + 0.01 \times (1 - 0) + 0.01 \times (1 - 0)\right)}$$

As you can see, the simple math behind Bayes' rule can quickly get tedious if you do it by hand. Fortunately for us, we live in the wonderous age of computers, so we can let Python handle the boring stuff!

Project #1: Search and Rescue

In this project, you'll write a Python program that uses Bayes' rule to find a solitary fisherman who has gone missing off Cape Python. As the director of the Coast Guard's search and rescue operations for the region, you've already interviewed his wife and determined his last known position, now more than six hours old. He radioed that he was abandoning ship, but no one knows if he is in a life raft or floating in the sea. The waters around the cape are warm, but if he's immersed, he'll experience hypothermia in 12 hours or so. If he's wearing a personal flotation device and lucky, he might last three days.

The ocean currents off Cape Python are complex (Figure 1-3), and the wind is currently blowing from the southwest. Visibility is good, but the waves are choppy, making a human head hard to spot.

Figure 1-3: Ocean currents off Cape Python

In real life, your next course of action would be to plug all the information you have into the Coast Guard's Search and Rescue Optimal Planning System (SAROPS). This software considers factors such as winds, tides, currents, whether a body is in the water or in a boat, and so on. It then generates rectangular search areas, calculates the initial probabilities for finding the sailor in each area, and plots the most efficient flight patterns.

For this project, you'll assume that SAROPS has identified three search areas. All you need to do is write the program that applies Bayes' rule. You also have enough resources available to search two of the three areas in a day. You'll have to decide how to allocate those resources. It's a lot of pressure, but you have a powerful assistant to help you out: Bayes' rule.

THE OBJECTIVE

Create a search and rescue game that uses Bayes' rule to inform player choices on how to conduct a search.

The Strategy

Searching for the sailor is like looking for the lost glasses in our previous example. You'll start with initial target probabilities for the sailor's location and update them for the search results. If you achieve an effective search of an area but find nothing, the probability that the sailor is in another area will increase.

But just as in real life, there are two ways things could go wrong: you thoroughly search an area but still miss the sailor, or your search goes poorly, wasting a day's effort. To equate this to search effectiveness scores, in the first case, you might get an SEP of 0.85, but the sailor is in the remaining 15 percent of the area not searched. In the second case, your SEP is 0.2, and you've left 80 percent of the area unsearched!

You can see the dilemma real commanders face. Do you go with your gut and ignore Bayes? Do you stick with the pure, cold logic of Bayes because you believe it's the best answer? Or do you act expediently and protect your career and reputation by going with Bayes even when you doubt it?

To aid the player, you'll use the OpenCV library to build an interface for working with the program. Although the interface can be something simple, like a menu built in the shell, you'll also want a map of the cape and the search areas. You'll use this map to display the sailor's last known position and his position when found. The OpenCV library is an excellent choice for this game since it lets you display images and add drawings and text.

Installing the Python Libraries

OpenCV is the world's most popular computer vision library. *Computer vision* is a field of deep learning that enables machines to see, identify, and process images like humans. OpenCV began as an Intel Research initiative in 1999 and is now maintained by the OpenCV Foundation, a nonprofit foundation which provides the software for free.

OpenCV is written in C++, but there are bindings in other languages, such as Python and Java. Although aimed primarily at real-time computer vision applications, OpenCV also includes common image manipulation tools such as those found in the Python Imaging Library. As of this writing, the current version is OpenCV 4.1.

OpenCV requires both the Numerical Python (NumPy) and SciPy packages to perform numerical and scientific computing in Python. OpenCV treats images as three-dimensional NumPy arrays (Figure 1-4). This allows for maximum interoperability with other Python scientific libraries.

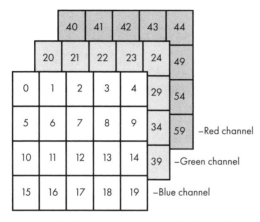

Figure 1-4: Visual representation of a three-channel color image array

OpenCV stores properties as rows, columns, and channels. For the image represented in Figure 1-4, its "shape" would be a three-element tuple (4, 5, 3). Each stack of cells, like 0-20-40 or 19-39-59, represents a single pixel. The numbers shown are the intensity values for each color channel for that pixel.

As many projects in this book require scientific Python libraries like NumPy and matplotlib, this is a good time to install them.

There are numerous ways to install these packages. One way is to use SciPy, an open source Python library used for scientific and technical computing (see *https://scipy.org/index.html*).

Alternatively, if you're going to do a lot of data analysis and plotting on your own time, you may want to download and use a free Python distribution like Anaconda or Enthought Canopy, which work with Windows, Linux, and macOS. These distributions spare you the task of finding and installing the correct versions of all the required data science libraries, such as NumPy, SciPy, and so on. A listing of these types of distributions, along with links to their websites, can be found at *https://scipy.org/install.html*.

Installing NumPy and Other Scientific Packages with pip

If you want to install the products directly, use the *Preferred Installer Program (pip)*, a package management system that makes it easy to install Python-based software (see *https://docs.python.org/3/installing/*). For Windows and macOS, Python versions 3.4 and newer come with pip preinstalled. Linux users may have to install pip separately. To install or upgrade pip, see the instructions at *https://pip.pypa.io/en/stable/installing/* or search online for instructions on installing pip on your particular operating system.

I used pip to install the scientific packages using the instructions at *https://scipy.org/install.html*. Because `matplotlib` requires multiple dependencies, you'll need to install these as well. For Windows, run the following Python 3–specific command using PowerShell, launched (using SHIFT-right-click) from within the folder containing the current Python installation:

```
$ python -m pip install --user numpy scipy matplotlib ipython jupyter pandas sympy nose
```

If you have both Python 2 and 3 installed, use `python3` in place of `python`.

To verify that `NumPy` has been installed and is available for OpenCV, open a Python shell and enter the following:

```
>>> import numpy
```

If you don't see an error, you're ready to install OpenCV.

Installing OpenCV with pip

You can find installation instructions for OpenCV at *https://pypi.org/project/opencv-python/*. To install OpenCV for standard desktop environments (Windows, macOS, and almost any GNU/Linux distribution), enter the following in a PowerShell or terminal window:

```
pip install opencv-contrib-python
```

or

```
python -m pip install opencv-contrib-python
```

If you have multiple versions of Python installed (such as versions 2.7 and 3.7), you will need to specify the Python version you want to use.

```
py -3.7 -m pip install --user opencv-contrib-python
```

If you're using Anaconda as a distribution medium, you can run this:

```
conda install opencv
```

To check that everything loaded properly, enter the following in the shell:

```
>>> import cv2
```

No error means you're good to go! If you get an error, read the trouble-shooting list at *https://pypi.org/project/opencv-python/*.

The Bayes Code

The *bayes.py* program you'll write in this section simulates the search for a missing sailor over three contiguous search areas. It will display a map, print a menu of search choices for the user, randomly choose a location for the sailor, and either reveal the location if a search locates him or do a Bayesian update of the probabilities of finding the sailor for each search area. You can download the code, along with the map image (*cape_python.png*), from *https://nostarch.com/real-world-python/*.

Importing Modules

Listing 1-1 starts the *bayes.py* program by importing the required modules and assigning some constants. We'll look at what these modules do as we implement them in the code.

bayes.py, part 1
```
import sys
import random
import itertools
import numpy as np
import cv2 as cv

MAP_FILE = 'cape_python.png'

SA1_CORNERS = (130, 265, 180, 315)  # (UL-X, UL-Y, LR-X, LR-Y)
SA2_CORNERS = (80, 255, 130, 305)   # (UL-X, UL-Y, LR-X, LR-Y)
SA3_CORNERS = (105, 205, 155, 255)  # (UL-X, UL-Y, LR-X, LR-Y)
```

Listing 1-1: Importing modules and assigning constants used in the bayes.py *program*

When importing modules into a program, the preferred order is the Python Standard Library modules, followed by third-party modules, followed by user-defined modules. The sys module includes commands for the operating system, such as exiting. The random module lets you generate pseudorandom numbers. The itertools module helps you with looping. Finally, numpy and cv2 import NumPy and OpenCV, respectively. You can also assign shorthand names (np, cv) to reduce keystrokes later.

Next, assign some constants. As per the PEP8 Python style guide (*https://www.python.org/dev/peps/pep-0008/*), constant names should be all caps. This doesn't make the variables truly immutable, but it does alert other developers that they shouldn't change these variables.

The map you'll use for the fictional Cape Python area is an image file called *cape_python.png* (Figure 1-5). Assign this image file to a constant variable named MAP_FILE.

Figure 1-5: Grayscale base map of Cape Python (cape_python.png)

You'll draw the search areas on the image as rectangles. OpenCV will define each rectangle by the pixel number at the corner points, so assign a variable to hold these four points as a tuple. The required order is upper-left *x*, upper-left *y*, lower-right *x*, and lower-right *y*. Use SA in the variable name to represent "search area."

Defining the Search Class

A *class* is a data type in object-oriented programming (OOP). OOP is an alternative approach to functional/procedural programming. It's especially useful for large, complex programs, as it produces code that's easier to update, maintain, and reuse, while reducing code duplication. OOP is built around data structures known as *objects*, which consist of data, methods, and the interactions between them. As such, it works well with game programs, which typically use interacting objects, such as spaceships and asteroids.

A class is a template from which multiple objects can be created. For example, you could have a class that builds battleships in a World War II game. Each battleship would inherit certain consistent characteristics, such as tonnage, cruising speed, fuel level, damage level, weaponry, and so on. You could also give each battleship object unique characteristics, such as a different name. Once created, or *instantiated*, the individual characteristics of each battleship would begin to diverge depending on how much fuel the ships burn, how much damage they take, how much ammo they use, and so on.

In *bayes.py*, you'll use a class as a template to create a search and rescue mission that allows for three search areas. Listing 1-2 defines the Search class, which will act as a blueprint for your game.

```
class Search():
    """Bayesian Search & Rescue game with 3 search areas."""

    def __init__(self, name):
        self.name = name
❶      self.img = cv.imread(MAP_FILE, cv.IMREAD_COLOR)
        if self.img is None:
            print('Could not load map file {}'.format(MAP_FILE),
                file=sys.stderr)
            sys.exit(1)

❷      self.area_actual = 0
        self.sailor_actual = [0, 0] # As "local" coords within search area

❸      self.sa1 = self.img[SA1_CORNERS[1] : SA1_CORNERS[3],
                            SA1_CORNERS[0] : SA1_CORNERS[2]]

        self.sa2 = self.img[SA2_CORNERS[1] : SA2_CORNERS[3],
                            SA2_CORNERS[0] : SA2_CORNERS[2]]

        self.sa3 = self.img[SA3_CORNERS[1] : SA3_CORNERS[3],
                            SA3_CORNERS[0] : SA3_CORNERS[2]]

❹      self.p1 = 0.2
        self.p2 = 0.5
        self.p3 = 0.3

        self.sep1 = 0
        self.sep2 = 0
        self.sep3 = 0
```

Listing 1-2: Defining the Search class and __init__() method

Start by defining a class called Search. According to PEP8, the first letter of a class name should be capitalized.

Next, define a method that sets up the initial attribute values for your object. In OOP, an *attribute* is a named value associated with an object. If your object is a person, an attribute might be their weight or eye color. *Methods* are attributes that also happen to be functions, which are passed a reference to their instance when they run. The __init__() method is a special built-in function that Python automatically invokes as soon as a new object is created. It binds the attributes of each newly created instance of a class. In this case, you pass it two arguments: self and the name you want to use for your object.

The self parameter is a reference to the instance of the class that is being created, or that a method was invoked on, technically referred to as a *context* instance. For example, if you create a battleship named the *Missouri*, then for that object, self becomes Missouri, and you can call a method for that object, like one for firing the big guns, with dot notation: Missouri.fire_big_guns(). By giving objects unique names when they are instantiated, the scope of each object's attributes is kept separate from all others. This way, damage taken by one battleship isn't shared with the rest of the fleet.

It's good practice to list all the initial attribute values for an object under the __init__() method. This way, users can see all the key attributes of the object that will be used later in various methods, and your code will be more readable and updatable. In Listing 1-2, these are the self attributes, such as self.name.

Attributes assigned to self will also behave like global variables in procedural programming. Methods in the class will be able to access them directly, without the need for arguments. Because these attributes are "shielded" under the *class* umbrella, their use is not discouraged as with true global variables, which are assigned within the global scope and are modified within the local scope of individual functions.

Assign the MAP_FILE variable to the self.img attribute using OpenCV's imread() method ❶. The MAP_FILE image is grayscale, but you'll want to add some color to it during the search. So, use ImreadFlag, as cv.IMREAD_COLOR, to load the image in color mode. This will set up three color channels (B, G, R) for you to exploit later.

If the image file doesn't exist (or the user entered the wrong filename), OpenCV will throw a confusing error (NoneType object is not subscriptable). To handle this, use a conditional to check whether self.img is None. If it is, print an error message and then use the sys module to exit the program. Passing it an exit code of 1 indicates that the program terminated with an error. Setting file=stderr will result in the use of the standard "error red" text color in the Python interpreter window, though not in other windows such as PowerShell.

Next, assign two attributes for the sailor's actual location when found. The first will hold the number of the search area ❷ and the second the precise (*x, y*) location. The assigned values will be placeholders for now. Later, you'll define a method to randomly choose the final values. Note that you use a list for the location coordinates as you need a mutable container.

The map image is loaded as an *array*. An array is a fixed-size collection of objects of the same type. Arrays are memory-efficient containers that provide fast numerical operations and effectively use the addressing logic of computers. One concept that makes NumPy particularly powerful is *vectorization*, which replaces explicit loops with more efficient array expressions. Basically, operations occur on entire arrays rather than their individual elements. With NumPy, internal looping is directed to efficient C and Fortran functions that are faster than standard Python techniques.

So that you can work with local coordinates *within* a search area, you can create a subarray from the array ❸. Notice that this is done with indexing. You first provide the range from the upper-left *y* value to the lower-right *y* and then from the upper-left *x* to the lower-right *x*. This is a NumPy feature that takes some getting used to, especially since most of us are used to *x* coming before *y* in Cartesian coordinates.

Repeat the procedure for the next two search areas and then set the pre-search probabilities for finding the sailor in each of the search areas ❹. In real life, these would come from the SAROPS program. Of course, p1 represents area 1, p2 is for area 2, and so on. Finish with placeholder attributes for the SEP.

Drawing the Map

Inside the Search class, you'll use functionality within OpenCV to create a method that displays the base map. This map will include the search areas, a scale bar, and the sailor's last known position (Figure 1-6).

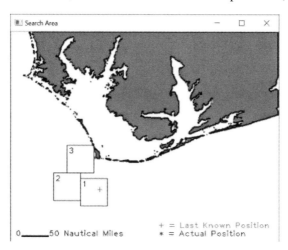

Figure 1-6: Initial game screen (base map) for bayes.py

Listing 1-3 defines the draw_map() method that displays the initial map.

bayes.py, part 3

```
def draw_map(self, last_known):
    """Display basemap with scale, last known xy location, search areas."""
    cv.line(self.img, (20, 370), (70, 370), (0, 0, 0), 2)
    cv.putText(self.img, '0', (8, 370), cv.FONT_HERSHEY_PLAIN, 1, (0, 0, 0))
    cv.putText(self.img, '50 Nautical Miles', (71, 370),
                cv.FONT_HERSHEY_PLAIN, 1, (0, 0, 0))

❶ cv.rectangle(self.img, (SA1_CORNERS[0], SA1_CORNERS[1]),
                        (SA1_CORNERS[2], SA1_CORNERS[3]), (0, 0, 0), 1)
    cv.putText(self.img, '1',
                (SA1_CORNERS[0] + 3, SA1_CORNERS[1] + 15),
                cv.FONT_HERSHEY_PLAIN, 1, 0)
    cv.rectangle(self.img, (SA2_CORNERS[0], SA2_CORNERS[1]),
                (SA2_CORNERS[2], SA2_CORNERS[3]), (0, 0, 0), 1)
    cv.putText(self.img, '2',
                (SA2_CORNERS[0] + 3, SA2_CORNERS[1] + 15),
                cv.FONT_HERSHEY_PLAIN, 1, 0)
    cv.rectangle(self.img, (SA3_CORNERS[0], SA3_CORNERS[1]),
                (SA3_CORNERS[2], SA3_CORNERS[3]), (0, 0, 0), 1)
    cv.putText(self.img, '3',
                (SA3_CORNERS[0] + 3, SA3_CORNERS[1] + 15),
                cv.FONT_HERSHEY_PLAIN, 1, 0)

❷ cv.putText(self.img, '+', (last_known),
                cv.FONT_HERSHEY_PLAIN, 1, (0, 0, 255))
    cv.putText(self.img, '+ = Last Known Position', (274, 355),
                cv.FONT_HERSHEY_PLAIN, 1, (0, 0, 255))
```

```
          cv.putText(self.img, '* = Actual Position', (275, 370),
                  cv.FONT_HERSHEY_PLAIN, 1, (255, 0, 0))

    ❸ cv.imshow('Search Area', self.img)
       cv.moveWindow('Search Area', 750, 10)
       cv.waitKey(500)
```

Listing 1-3: Defining a method for displaying the base map

Define the draw_map() method with self and the sailor's last known coordinates (last_known) as its two parameters. Then use OpenCV's line() method to draw a scale bar. Pass it the base map image, a tuple of the left and right (*x*, *y*) coordinates, a line color tuple, and a line width as arguments.

Use the putText() method to annotate the scale bar. Pass it the attribute for the base map image and then the actual text, followed by a tuple of the coordinates of the bottom-left corner of the text. Then add the font name, font scale, and color tuple.

Now draw a rectangle for the first search area ❶. As usual, pass the base map image, then the variables representing the four corners of the box, and finally a color tuple and a line weight. Use putText() again to place the search area number just inside the upper-left corner. Repeat these steps for search areas 2 and 3.

Use putText() to post a + at the sailor's last known position ❷. Note that the symbol is red, but the color tuple reads (0, 0, 255), instead of (255, 0, 0). This is because OpenCV uses a Blue-Green-Red (BGR) color format, not the more common Red-Green-Blue (RGB) format.

Continue by placing text for a legend that describes the symbols for the last known position and actual position, which should display when a player's search finds the sailor. Use blue for the actual position marker.

Complete the method by showing the base map, using OpenCV's imshow() method ❸. Pass it a title for the window and the image.

To avoid the base map and interpreter windows interfering with each other as much as possible, force the base map to display in the upper-right corner of your monitor (you may need to adjust the coordinates for your machine). Use OpenCV's moveWindow() method and pass it the name of the window, 'Search Area', and the coordinates for the top-left corner.

Finish by using the waitKey() method, which introduces a delay of *n* milliseconds while rendering images to windows. Pass it 500, for 500 milliseconds. This should result in the game menu appearing a half-second after the base map.

Choosing the Sailor's Final Location

Listing 1-4 defines a method to randomly choose the sailor's actual location. For convenience, the coordinates are initially found within a search area subarray and then converted to global coordinates with respect to the full base map image. This methodology works because all the search areas are the same size and shape and can thus use the same internal coordinates.

```
def sailor_final_location(self, num_search_areas):
    """Return the actual x,y location of the missing sailor."""
    # Find sailor coordinates with respect to any Search Area subarray.
    self.sailor_actual[0] = np.random.choice(self.sa1.shape[1], 1)
    self.sailor_actual[1] = np.random.choice(self.sa1.shape[0], 1)

  ❶ area = int(random.triangular(1, num_search_areas + 1))

    if area == 1:
        x = self.sailor_actual[0] + SA1_CORNERS[0]
        y = self.sailor_actual[1] + SA1_CORNERS[1]
      ❷ self.area_actual = 1
    elif area == 2:
        x = self.sailor_actual[0] + SA2_CORNERS[0]
        y = self.sailor_actual[1] + SA2_CORNERS[1]
        self.area_actual = 2
    elif area == 3:
        x = self.sailor_actual[0] + SA3_CORNERS[0]
        y = self.sailor_actual[1] + SA3_CORNERS[1]
        self.area_actual = 3
    return x, y
```

Listing 1-4: Defining a method to randomly choose the sailor's actual location

Define the sailor_final_location() method with two parameters: self and the number of search areas being used. For the first (x) coordinate in the self.sailor_actual list, use NumPy's random.choice() method to choose a value from the area 1 subarray. Remember, the search areas are NumPy arrays copied out of the larger image array. Because the search areas/subarrays are all the same size, coordinates you choose from one will apply to all.

You can get the coordinates of an array with shape, as shown here:

```
>>> print(np.shape(self.SA1))
(50, 50, 3)
```

The shape attribute for a NumPy array must be a tuple with as many elements as dimensions in the array. And remember that, for an array in OpenCV, the order of elements in the tuple is rows, columns, and then channels.

Each of the existing search areas is a three-dimensional array 50×50 pixels in size. So, internal coordinates for both x and y will range from 0 to 49. Selecting [0] with random.choice() means that rows are used, and the final argument, 1, selects a single element. Selecting [1] chooses from columns.

The coordinates generated by random.choice() will range from 0 to 49. To use these with the full base map image, you first need to pick a search area ❶. Do this with the random module, which you imported at the start of the program. According to the SAROPS output, the sailor is most likely in area 2, followed by area 3. Since these initial target probabilities are guesses that won't correspond directly to reality, use a triangular distribution to choose the area containing the sailor. The arguments are the low and high endpoints. If a final mode argument is not provided, the mode defaults to the midpoint between the endpoints. This will align with the SAROPS results as area 2 will be picked the most often.

Note that you use the local variable area within the method, rather than the self.area attribute, as there's no need to share this variable with other methods.

To plot the sailor's location on the base map, you need to add the appropriate search area corner-point coordinate. This converts the "local" search area coordinates to the "global" coordinates of the full base map image. You'll also want to keep track of the search area, so update the self.area_actual attribute ❷.

Repeat these steps for search areas 2 and 3 and then return the (x, y) coordinates.

In real life, the sailor would drift along, and the odds of his moving into area 3 would increase with each search. I chose to use a static location, however, to make the logic behind Bayes' rule as clear as possible. As a result, this scenario behaves more like a search for a sunken submarine.

Calculating Search Effectiveness and Conducting the Search

In real life, weather and mechanical problems can result in low search effectiveness scores. Thus, the strategy for each search will be to generate a list of all possible locations within a search area, shuffle the list, and then sample it based on the search effectiveness value. Because the SEP will never be 1.0, if you just sample from the start or end of the list—without shuffling—you'll never be able to access coordinates tucked away in its "tail."

Listing 1-5, still in the Search class, defines a method to randomly calculate the effectiveness of a given search and defines another method to conduct the search.

bayes.py, part 5

```
def calc_search_effectiveness(self):
    """Set decimal search effectiveness value per search area."""
    self.sep1 = random.uniform(0.2, 0.9)
    self.sep2 = random.uniform(0.2, 0.9)
    self.sep3 = random.uniform(0.2, 0.9)

❶ def conduct_search(self, area_num, area_array, effectiveness_prob):
    """Return search results and list of searched coordinates."""
    local_y_range = range(area_array.shape[0])
    local_x_range = range(area_array.shape[1])
❷   coords = list(itertools.product(local_x_range, local_y_range))
    random.shuffle(coords)
    coords = coords[:int((len(coords) * effectiveness_prob))]
❸   loc_actual = (self.sailor_actual[0], self.sailor_actual[1])
    if area_num == self.area_actual and loc_actual in coords:
        return 'Found in Area {}.'.format(area_num), coords
    else:
        return 'Not Found', coords
```

Listing 1-5: Defining methods to randomly choose search effectiveness and conduct search

Start by defining the search effectiveness method. The only parameter needed is self. For each of the search effectiveness attributes, such as E1, randomly choose a value between 0.2 and 0.9. These are arbitrary values that mean you will always search at least 20 percent of the area but never more than 90 percent.

You could argue that the search effectiveness attributes for the three search areas are dependent. Fog, for example, might affect all three areas, yielding uniformly poor results. On the other hand, some of your helicopters may have infrared imaging equipment and would fare better. At any rate, making these independent, as you've done here, makes for a more dynamic simulation.

Next, define a method for conducting a search ❶. Necessary parameters are the object itself, the area number (chosen by the user), the subarray for the chosen area, and the randomly chosen search effectiveness value.

You'll need to generate a list of all the coordinates within a given search area. Name a variable local_y_range and assign it a range based on the first index from the array shape tuple, which represents rows. Repeat for the x_range value.

To generate the list of all coordinates in the search area, use the itertools module ❷. This module is a group of functions in the Python Standard Library that create iterators for efficient looping. The product() function returns tuples of all the permutations-with-repetition for a given sequence. In this case, you're finding all the possible ways to combine x and y in the search area. To see it in action, type the following in the shell:

```
>>> import itertools
>>> x_range = [1, 2, 3]
>>> y_range = [4, 5, 6]
>>> coords = list(itertools.product(x_range, y_range))
>>> coords
[(1, 4), (1, 5), (1, 6), (2, 4), (2, 5), (2, 6), (3, 4), (3, 5), (3, 6)]
```

As you can see, the coords list contains every possible paired combination of the elements in the x_range and y_range lists.

Next, shuffle the list of coordinates. This is so you won't keep searching the same end of the list with each search event. In the next line, use index slicing to trim the list based on the search effectiveness probability. For example, a poor search effectiveness of 0.3 means that only one-third of the possible locations in an area are included in the list. As you'll check the sailor's actual location against this list, you'll effectively leave two-thirds of the area "unsearched."

Assign a local variable, loc_actual, to hold the sailor's actual location ❸. Then use a conditional to check that the sailor has been found. If the user chose the correct search area and the shuffled and trimmed coords list contains the sailor's (x, y) location, return a string stating the sailor has been found, along with the coords list. Otherwise, return a string stating the sailor has not been found and the coords list.

Applying Bayes' Rule and Drawing a Menu

Listing 1-6, still in the Search class, defines a method and a function. The revise_target_probs() method uses Bayes' rule to update the target probabilities. These represent the probability of the sailor being found per search area. The draw_menu() function, defined outside of the Search class, displays a menu that will serve as a graphical user interface (GUI) to run the game.

bayes.py, part 6

```python
def revise_target_probs(self):
    """Update area target probabilities based on search effectiveness."""
    denom = self.p1 * (1 - self.sep1) + self.p2 * (1 - self.sep2) \
            + self.p3 * (1 - self.sep3)
    self.p1 = self.p1 * (1 - self.sep1) / denom
    self.p2 = self.p2 * (1 - self.sep2) / denom
    self.p3 = self.p3 * (1 - self.sep3) / denom


def draw_menu(search_num):
    """Print menu of choices for conducting area searches."""
    print('\nSearch {}'.format(search_num))
    print(
        """
        Choose next areas to search:

        0 - Quit
        1 - Search Area 1 twice
        2 - Search Area 2 twice
        3 - Search Area 3 twice
        4 - Search Areas 1 & 2
        5 - Search Areas 1 & 3
        6 - Search Areas 2 & 3
        7 - Start Over
        """
    )
```

Listing 1-6: Defining ways to apply Bayes' rule and draw a menu in the Python shell

Define the revise_target_probs() method to update the probability of the sailor being in each search area. Its only parameter is self.

For convenience, break Bayes' equation into two parts, starting with the denominator. You need to multiply the previous target probability by the current search effectiveness value (see page 5 to review how this works).

With the denominator calculated, use it to complete Bayes' equation. In OOP, you don't need to return anything. You can simply update the attribute directly in the method, as if it were a declared global variable in procedural programming.

Next, in the global space, define the draw_menu() function to draw a menu. Its only parameter is the number of the search being conducted. Because this function has no "self-use," you don't have to include it in the class definition, though that is a valid option.

Start by printing the search number. You'll need this to keep track of whether you've found the sailor in the requisite number of searches, which we've currently set as 3.

Use triple quotes with the print() function to display the menu. Note that the user will have the option to allocate both search parties to a given area or divide them between two areas.

Defining the main() Function

Now that you're finished with the Search class, you're ready to put all those attributes and methods to work! Listing 1-7 begins the definition of the main() function, used to run the program.

bayes.py, part 7

```python
def main():
    app = Search('Cape_Python')
    app.draw_map(last_known=(160, 290))
    sailor_x, sailor_y = app.sailor_final_location(num_search_areas=3)
    print("-" * 65)
    print("\nInitial Target (P) Probabilities:")
    print("P1 = {:.3f}, P2 = {:.3f}, P3 = {:.3f}".format(app.p1, app.p2, app.p3))
    search_num = 1
```

Listing 1-7: Defining the start of the main() function, used to run the program

The main() function requires no arguments. Start by creating a game application, named app, using the Search class. Name the object Cape_Python.

Next, call the method that displays the map. Pass it the last known position of the sailor as a tuple of (*x*, *y*) coordinates. Note the use of the keyword argument, last_known=(160, 290), for clarity.

Now, get the sailor's *x* and *y* location by calling the method for that task and passing it the number of search areas. Then print the initial target probabilities, or priors, which were calculated by your Coast Guard underlings using Monte Carlo simulation, not Bayes' rule. Finally, name a variable search_num and assign it 1. This variable will keep track of how many searches you've conducted.

Evaluating the Menu Choices

Listing 1-8 starts the while loop used to run the game in main(). Within this loop, the player evaluates and selects menu choices. Choices include searching a single area twice, splitting search efforts between two areas, restarting the game, and exiting the game. Note that the player can conduct as many searches as it takes to find the sailor; our three-day limit hasn't been "hard-wired" into the game.

bayes.py, part 8

```python
    while True:
        app.calc_search_effectiveness()
        draw_menu(search_num)
        choice = input("Choice: ")

        if choice == "0":
            sys.exit()
```

```
❶ elif choice == "1":
      results_1, coords_1 = app.conduct_search(1, app.sa1, app.sep1)
      results_2, coords_2 = app.conduct_search(1, app.sa1, app.sep1)
   ❷ app.sep1 = (len(set(coords_1 + coords_2))) / (len(app.sa1)**2)
      app.sep2 = 0
      app.sep3 = 0

   elif choice == "2":
      results_1, coords_1 = app.conduct_search(2, app.sa2, app.sep2)
      results_2, coords_2 = app.conduct_search(2, app.sa2, app.sep2)
      app.sep1 = 0
      app.sep2 = (len(set(coords_1 + coords_2))) / (len(app.sa2)**2)
      app.sep3 = 0

   elif choice == "3":
      results_1, coords_1 = app.conduct_search(3, app.sa3, app.sep3)
      results_2, coords_2 = app.conduct_search(3, app.sa3, app.sep3)
      app.sep1 = 0
      app.sep2 = 0
      app.sep3 = (len(set(coords_1 + coords_2))) / (len(app.sa3)**2)

❸ elif choice == "4":
      results_1, coords_1 = app.conduct_search(1, app.sa1, app.sep1)
      results_2, coords_2 = app.conduct_search(2, app.sa2, app.sep2)
      app.sep3 = 0

   elif choice == "5":
      results_1, coords_1 = app.conduct_search(1, app.sa1, app.sep1)
      results_2, coords_2 = app.conduct_search(3, app.sa3, app.sep3)
      app.sep2 = 0

   elif choice == "6":
      results_1, coords_1 = app.conduct_search(2, app.sa2, app.sep2)
      results_2, coords_2 = app.conduct_search(3, app.sa3, app.sep3)
      app.sep1 = 0

❹ elif choice == "7":
      main()

   else:
      print("\nSorry, but that isn't a valid choice.", file=sys.stderr)
      continue
```

Listing 1-8: Using a loop to evaluate menu choices and run the game

Start a while loop that will run until the user chooses to exit. Immediately use dot notation to call the method that calculates the effectiveness of the search. Then call the function that displays the game menu and pass it the search number. Finish the preparatory stage by asking the user to make a choice, using the input() function.

The player's choice will be evaluated using a series of conditional statements. If they choose 0, exit the game. Exiting uses the sys module you imported at the beginning of the program.

If the player chooses 1, 2, or 3, it means they want to commit both search teams to the area with the corresponding number. You'll need to call the conduct_search() method twice to generate two sets of results and coordinates ❶. The tricky part here is determining the overall SEP, since each search has its own SEP. To do this, add the two coords lists together and convert the result to a set to remove any duplicates ❷. Get the length of the set and then divide it by the number of pixels in the 50×50 search area. Since you didn't search the other areas, set their SEPs to 0.

Repeat and tailor the previous code for search areas 2 and 3. Use an elif statement since only one menu choice is valid per loop. This is more efficient than using additional if statements, as all elif statements below a true response will be skipped.

If the player chooses a 4, 5, or 6, it means they want to divide their teams between two areas. In this case, there's no need to recalculate the SEP ❸.

If the player finds the sailor and wants to play again or just wants to restart, call the main() function ❹. This will reset the game and clear the map.

If the player makes a nonvalid choice, like "Bob", let them know with a message and then use continue to skip back to the start of the loop and request the player's choice again.

Finishing and Calling main()

Listing 1-9, still in the while loop, finishes the main() function and then calls it to run the program.

bayes.py, part 9

```
        app.revise_target_probs()  # Use Bayes' rule to update target probs.

        print("\nSearch {} Results 1 = {}"
              .format(search_num, results_1), file=sys.stderr)
        print("Search {} Results 2 = {}\n"
              .format(search_num, results_2), file=sys.stderr)
        print("Search {} Effectiveness (E):".format(search_num))
        print("E1 = {:.3f}, E2 = {:.3f}, E3 = {:.3f}"
              .format(app.sep1, app.sep2, app.sep3))

❶      if results_1 == 'Not Found' and results_2 == 'Not Found':
            print("\nNew Target Probabilities (P) for Search {}:"
                  .format(search_num + 1))
            print("P1 = {:.3f}, P2 = {:.3f}, P3 = {:.3f}"
                  .format(app.p1, app.p2, app.p3))
        else:
            cv.circle(app.img, (sailor_x, sailor_y), 3, (255, 0, 0), -1)
❷          cv.imshow('Search Area', app.img)
            cv.waitKey(1500)
            main()
        search_num += 1

if __name__ == '__main__':
    main()
```

Listing 1-9: Completing and calling the main() function

Call the `revise_target_probs()` method to apply Bayes' rule and recalculate the probability of the sailor being in each search area, given the search results. Next, display the search results and search effectiveness probabilities in the shell.

If the results of both searches are negative, display the updated target probabilities, which the player will use to guide their next search ❶. Otherwise, display the sailor's location on the map. Use OpenCV to draw a circle and pass the method the base map image, the sailor's (*x*, *y*) tuple for the center point, a radius (in pixels), a color, and a thickness of –1. A negative thickness value will fill the circle with the color.

Finish `main()` by showing the base map using code similar to Listing 1-3 ❷. Pass the `waitKey()` method 1500 to display the sailor's actual location for 1.5 seconds before the game calls `main()` and resets automatically. At the end of the loop, increment the search number variable by 1. You want to do this *after* the loop so that an invalid choice isn't counted as a search.

Back in the global space, apply the code that lets the program be imported as a module or run in stand-alone mode. The `__name__` variable is a built-in variable used to evaluate whether a program is autonomous or imported into another program. If you run this program directly, `__name__` is set to `__main__`, the condition of the `if` statement is met, and `main()` is called automatically. If the program is imported, the `main()` function won't be run until it is intentionally called.

Playing the Game

To play the game, select **Run ▸ Run Module** in the text editor or just press F5. Figures 1-7 and 1-8 show the final game screens, with the results of a successful first search.

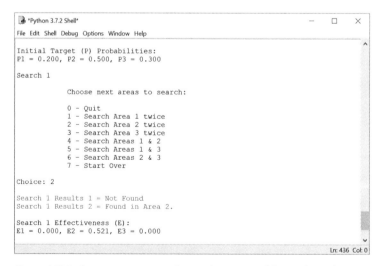

Figure 1-7: Python interpreter window with a successful search result

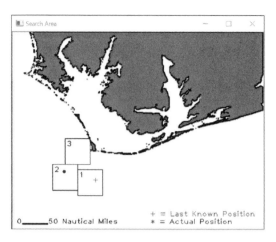

Figure 1-8: Base map image for a successful search result

In this example search, the player chose to commit both searches to area 2, which had an initial 50 percent probability of containing the sailor. The first search was unsuccessful, but the second one found the sailor. Note that the search effectiveness was only slightly better than 50 percent. This means there was only a one-in-four chance ($0.5 \times 0.521 = 0.260$) of finding the sailor in the first search. Despite choosing wisely, the player still had to rely on a bit of luck in the end!

When you play the game, try to immerse yourself in the scenario. Your decisions determine whether a human being lives or dies, and you don't have much time. If the sailor's floating in the water, you've got only three guesses to get it right. Use them wisely!

Based on the target probabilities at the start of the game, the sailor is most likely in area 2, followed by area 3. So, a good initial strategy is to either search area 2 twice (menu option 2) or search areas 2 and 3 simultaneously (menu option 6). You'll want to keep a close eye on the search effectiveness output. If an area gets a high effectiveness score, which means that it's been thoroughly searched, you may want to focus your efforts elsewhere for the rest of the game.

The following output represents one of the worst situations you can find yourself in as a decision maker:

```
Search 2 Results 1 = Not Found
Search 2 Results 2 = Not Found

Search 2 Effectiveness (E):
E1 = 0.000, E2 = 0.234, E3 = 0.610

New Target Probabilities (P) for Search 3:
P1 = 0.382, P2 = 0.395, P3 = 0.223
```

After search 2, with only one search left, the target probabilities are so similar they provide little guidance for where to search next. In this case, it's best to divide your searches between two areas and hope for the best.

Play the game a few times by blindly searching the areas in order of initial probability, doubling up on area 2, then 3, then 1. Then try obeying the Bayes results religiously, always doubling your searches in the area with the highest current target probability. Next, try dividing your searches between the areas with the two highest probabilities. After that, allow your own intuition to have a say, overruling Bayes when you feel it's appropriate. As you can imagine, with more search areas and more search days, human intuition would quickly get overwhelmed.

Summary

In this chapter, you learned about Bayes' rule, a simple statistical theorem with broad applications in our modern world. You wrote a program that used the rule to take new information—in the form of estimates of search effectiveness—and update the probability of finding a lost sailor in each area being searched.

You also loaded and used multiple scientific packages, like NumPy and OpenCV, that you'll implement throughout the book. And you applied the useful itertools, sys, and random modules from the Python Standard Library.

Further Reading

The Theory That Would Not Die: How Bayes' Rule Cracked the Enigma Code, Hunted Down Russian Submarines, and Emerged Triumphant from Two Centuries of Controversy (Yale University Press, 2011), by Sharon Bertsch McGrayne, recounts the discovery and controversial history of Bayes' rule. The appendix includes several example applications of Bayes' rule, one of which inspired the missing-sailor scenario used in this chapter.

A major source of documentation for NumPy is *https://docs.scipy.org/doc/*.

Challenge Project: Smarter Searches

Currently, the *bayes.py* program places all the coordinates within a search area into a list and randomly shuffles them. Subsequent searches in the same area may end up retracing previous tracks. This isn't necessarily bad from a real-life perspective, as the sailor will be drifting around the whole time, but overall it would be best to cover as much of the area as possible without repetition.

Copy and edit the program so that it keeps track of which coordinates have been searched within an area and excludes them from future searches (until main() is called again, either because the player finds the sailor or chooses menu option 7 to restart). Test the two versions of the game to see whether your changes noticeably impact the results.

Challenge Project: Finding the Best Strategy with MCS

Monte Carlo simulation (MCS) uses repeated random sampling to predict different outcomes under a specified range of conditions. Create a version of *bayes.py* that automatically chooses menu items and keeps track of thousands of results, allowing you to determine the most successful search strategy. For example, have the program choose menu item 1, 2, or 3 based on the highest Bayesian target probability and then record the search number when the sailor is found. Repeat this procedure 10,000 times and take the average of all the search numbers. Then loop again, choosing from menu item 4, 5, or 6 based on the highest combined target probability. Compare the final averages. Is it better to double up your searches in a single area or split them between two areas?

Challenge Project: Calculating the Probability of Detection

In a real-life search and rescue operation, you would make an estimate of the *expected* search effectiveness probability for each area prior to making a search. This expected, or *planned*, probability would be informed primarily by weather reports. For example, fog might roll into one search area, while the other two enjoy clear skies.

Multiplying target probability by the planned SEP yields the *probability of detection (PoD)* for an area. The PoD is the probability an object will be detected given all known error and noise sources.

Write a version of *bayes.py* that includes a randomly generated planned SEP for each search area. Multiply the target probability for each area (such as self.p1, self.p2, or self.p3) by these new variables to produce a PoD for the area. For example, if the Bayes target probability for area 3 is 0.90 but the planned SEP is only 0.1, then the probability of detection is 0.09.

In the shell display, show the player the target probabilities, the planned SEPs, and the PoD for each area, as shown next. Players can then use this information to guide their choice from the search menu.

```
Actual Search 1 Effectiveness (E):
E1 = 0.190, E2 = 0.000, E3 = 0.000

New Planned Search Effectiveness and Target Probabilities (P) for Search 2:
E1 = 0.509, E2 = 0.826, E3 = 0.686
P1 = 0.168, P2 = 0.520, P3 = 0.312

Search 2

    Choose next areas to search:

    0 - Quit

    1 - Search Area 1 twice
        Probability of detection: 0.164
```

```
   2 - Search Area 2 twice
       Probability of detection: 0.674

   3 - Search Area 3 twice
       Probability of detection: 0.382

   4 - Search Areas 1 & 2
       Probability of detection: 0.515

   5 - Search Areas 1 & 3
       Probability of detection: 0.3

   6 - Search Areas 2 & 3
       Probability of detection: 0.643

   7 - Start Over

Choice:
```

To combine PoD when searching the same area twice, use this formula:

$$1 - (1 - PoD)^2$$

Otherwise, just sum the probabilities.

When calculating the actual SEP for an area, constrain it somewhat to the expected value. This considers the general accuracy of weather reports made only a day in advance. Replace the `random.uniform()` method with a distribution, such as triangular, built around the planned SEP value. For a list of available distribution types, see *https://docs.python.org/3/library/random.html#real-valued-distributions*. Of course, the actual SEP for an unsearched area will always be zero.

How does incorporating planned SEPs affect gameplay? Is it easier or harder to win? Is it harder to grasp how Bayes' rule is being applied? If you oversaw a real search, how would you deal with an area with a high target probability but a low planned SEP due to rough seas? Would you search anyway, call off the search, or move the search to an area with a low target probability but better weather?

2

ATTRIBUTING AUTHORSHIP
WITH STYLOMETRY

Stylometry is the quantitative study of literary style through computational text analysis. It's based on the idea that we all have a unique, consistent, and recognizable style to our writing. This includes our vocabulary, our use of punctuation, the average length of our sentences and words, and so on.

A common application of stylometry is authorship attribution. Do you ever wonder if Shakespeare really wrote all his plays? Or if John Lennon or Paul McCartney wrote the song "In My Life"? Could Robert Galbraith, author of *A Cuckoo's Calling*, really be J. K. Rowling in disguise? Stylometry can find the answer!

Stylometry has been used to overturn murder convictions and even helped identify and convict the Unabomber in 1996. Other uses include detecting plagiarism and determining the emotional tone behind words, such as in social media posts. Stylometry can even be used to detect signs of mental depression and suicidal tendencies.

In this chapter, you'll use multiple stylometric techniques to determine whether Sir Arthur Conan Doyle or H. G. Wells wrote the novel *The Lost World*.

Project #2: The Hound, The War, and The Lost World

Sir Arthur Conan Doyle (1859–1930) is best known for the Sherlock Holmes stories, considered milestones in the field of crime fiction. H. G. Wells (1866–1946) is famous for several groundbreaking science fiction novels including *The War of The Worlds*, *The Time Machine*, *The Invisible Man*, and *The Island of Dr. Moreau*.

In 1912, the *Strand Magazine* published *The Lost World*, a serialized version of a science fiction novel. It told the story of an Amazon basin expedition, led by zoology professor George Edward Challenger, that encountered living dinosaurs and a vicious tribe of ape-like creatures.

Although the author of the novel is known, for this project, let's pretend it's in dispute and it's your job to solve the mystery. Experts have narrowed the field down to two authors, Doyle and Wells. Wells is slightly favored because *The Lost World* is a work of science fiction, which is his purview. It also includes brutish troglodytes redolent of the morlocks in his 1895 work *The Time Machine*. Doyle, on the other hand, is known for detective stories and historical fiction.

THE OBJECTIVE

Write a Python program that uses stylometry to determine whether Sir Arthur Conan Doyle or H. G. Wells wrote the novel *The Lost World*.

The Strategy

The science of *natural language processing (NLP)* deals with the interactions between the precise and structured language of computers and the nuanced, frequently ambiguous "natural" language used by humans. Example uses for NLP include machine translations, spam detection, comprehension of search engine questions, and predictive text recognition for cell phone users.

The most common NLP tests for authorship analyze the following features of a text:

Word length   A frequency distribution plot of the length of words in a document

Stop words   A frequency distribution plot of stop words (short, noncontextual function words like *the*, *but*, and *if*)

Parts of speech   A frequency distribution plot of words based on their syntactic functions (such as nouns, pronouns, verbs, adverbs, adjectives, and so on)

Most common words   A comparison of the most commonly used words in a text

Jaccard similarity   A statistic used for gauging the similarity and diversity of a sample set

If Doyle and Wells have distinctive writing styles, these five tests should be enough to distinguish between them. We'll talk about each test in more detail in the coding section.

To capture and analyze each author's style, you'll need a representative *corpus*, or a body of text. For Doyle, use the famous Sherlock Holmes novel *The Hound of the Baskervilles*, published in 1902. For Wells, use *The War of the Worlds*, published in 1898. Both these novels contain more than 50,000 words, more than enough for a sound statistical sampling. You'll then compare each author's sample to *The Lost World* to determine how closely the writing styles match.

To perform stylometry, you'll use the *Natural Language Toolkit (NLTK)*, a popular suite of programs and libraries for working with human language data in Python. It's free and works on Windows, macOS, and Linux. Created in 2001 as part of a computational linguistics course at the University of Pennsylvania, NLTK has continued to develop and expand with the help of dozens of contributors. To learn more, check out the official NLTK website at *http://www.nltk.org/*.

Installing NLTK

You can find installation instructions for NLTK at *http://www.nltk.org/install .html*. To install NLTK on Windows, open PowerShell and install it with Preferred Installer Program (pip).

```
python -m pip install nltk
```

If you have multiple versions of Python installed, you'll need to specify the version. Here's the command for Python 3.7:

```
py -3.7 -m pip install nltk
```

To check that the installation was successful, open the Python interactive shell and enter the following:

```
>>> import nltk
>>>
```

If you don't get an error, you're good to go. Otherwise, follow the installation instructions at *http://www.nltk.org/install.html*.

Downloading the Tokenizer

To run the stylometric tests, you'll need to break the multiple texts—or *corpora*—into individual words, referred to as *tokens*. At the time of this writing, the word_tokenize() method in NLTK implicitly calls sent_tokenize(), used to break a corpus into individual sentences. For handling sent_tokenize(), you'll need the *Punkt Tokenizer Models*. Although this is part of NLTK, you'll have to download it separately with the handy NLTK Downloader. To launch it, enter the following into the Python shell:

```
>>> import nltk
>>> nltk.download()
```

The NLTK Downloader window should now be open (Figure 2-1). Click either the **Models** or **All Packages** tab near the top; then click **punkt** in the Identifier column. Scroll to the bottom of the window and set the Download Directory for your platform (see *https://www.nltk.org/data.html*). Finally, click the **Download** button to download the Punkt Tokenizer Models.

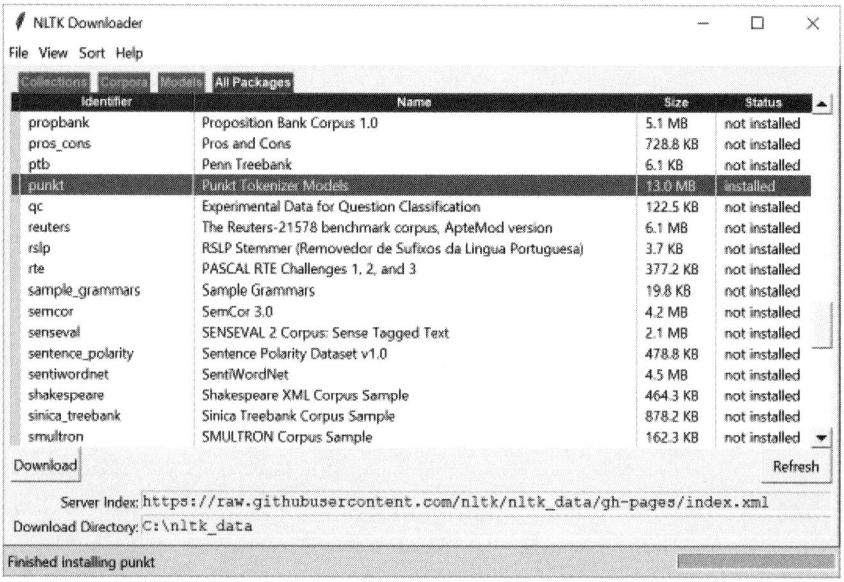

Figure 2-1: Downloading the Punkt Tokenizer Models

Note that you can also download NLTK packages directly in the shell. Here's an example:

```
>>> import nltk
>>> nltk.download('punkt')
```

You'll also need access to the Stopwords Corpus, which can be downloaded in a similar manner.

Downloading the Stopwords Corpus

Click the **Corpora** tab in the NLTK Downloader window and download the Stopwords Corpus, as shown in Figure 2-2.

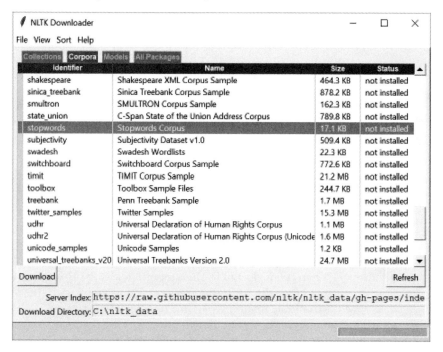

Figure 2-2: Downloading the Stopwords Corpus

Alternatively, you can use the shell.

```
>>> import nltk
>>> nltk.download('stopwords')
```

Let's download one more package to help you analyze parts of speech, like nouns and verbs. Click the **All Packages** tab in the NLTK Downloader window and download the Averaged Perceptron Tagger.

To use the shell, enter the following:

```
>>> import nltk
>>> nltk.download('averaged_perceptron_tagger')
```

When NLTK has finished downloading, exit the NLTK Downloader window and enter the following into the Python interactive shell:

```
>>> from nltk import punkt
```

Then enter the following:

```
>>> from nltk.corpus import stopwords
```

If you don't encounter an error, the models and corpus successfully downloaded.

Finally, you'll need `matplotlib` to make plots. If you haven't installed it already, see the instructions for installing scientific packages on page 6.

The Corpora

You can download the text files for *The Hound of the Baskervilles* (*hound.txt*), *The War of the Worlds* (*war.txt*), and *The Lost World* (*lost.txt*), along with the book's code, from *https://nostarch.com/real-world-python/*.

These came from Project Gutenberg (*http://www.gutenberg.org/*), a great source for public domain literature. So that you can use these texts right away, I've stripped them of extraneous material such as table of contents, chapter titles, copyright information, and so on.

The Stylometry Code

The *stylometry.py* program you'll write next loads the text files as strings, tokenizes them into words, and then runs the five stylometric analyses listed on pages 28–29. The program will output a combination of plots and shell messages that will help you determine who wrote *The Lost World*.

Keep the program in the same folder as the three text files. If you don't want to enter the code yourself, just follow along with the downloadable code available at *https://nostarch.com/real-world-python/*.

Importing Modules and Defining the main() Function

Listing 2-1 imports NLTK and `matplotlib`, assigns a constant, and defines the `main()` function to run the program. The functions used in `main()` will be described in detail later in the chapter.

stylometry.py, part 1

```
import nltk
from nltk.corpus import stopwords
import matplotlib.pyplot as plt

LINES = ['-', ':', '--']  # Line style for plots.

def main():
❶   strings_by_author = dict()
    strings_by_author['doyle'] = text_to_string('hound.txt')
    strings_by_author['wells'] = text_to_string('war.txt')
    strings_by_author['unknown'] = text_to_string('lost.txt')

    print(strings_by_author['doyle'][:300])

❷   words_by_author = make_word_dict(strings_by_author)
    len_shortest_corpus = find_shortest_corpus(words_by_author)
❸   word_length_test(words_by_author, len_shortest_corpus)
    stopwords_test(words_by_author, len_shortest_corpus)
    parts_of_speech_test(words_by_author, len_shortest_corpus)
```

```
    vocab_test(words_by_author)
    jaccard_test(words_by_author, len_shortest_corpus)
```

Listing 2-1: Importing modules and defining the main() function

Start by importing NLTK and the Stopwords Corpus. Then import `matplotlib`.

Create a variable called `LINES` and use the all-caps convention to indicate it should be treated as a constant. By default, `matplotlib` plots in color, but you'll still want to designate a list of symbols for color-blind people and this black-and-white book!

Define `main()` at the start of the program. The steps in this function are almost as readable as pseudocode and provide a good overview of what the program will do. The first step will be to initialize a dictionary to hold the text for each author ❶. The `text_to_string()` function will load each corpus into this dictionary as a string. The name of each author will be the dictionary key (using unknown for *The Lost World*), and the string of text from their novel will be the value. For example, here's the key, `Doyle`, with the value text string greatly truncated:

```
{'Doyle': 'Mr. Sherlock Holmes, who was usually very late in the mornings --snip--'}
```

Immediately after populating the dictionary, print the first 300 items for the doyle key to ensure things went as planned. This should produce the following printout:

```
Mr. Sherlock Holmes, who was usually very late in the mornings, save
upon those not infrequent occasions when he was up all night, was seated
at the breakfast table. I stood upon the hearth-rug and picked up the
stick which our visitor had left behind him the night before. It was a
fine, thick piec
```

With the corpora loaded correctly, the next step is to tokenize the strings into words. Currently, Python doesn't recognize words but instead works on *characters*, such as letters, numbers, and punctuation marks. To remedy this, you'll use the `make_word_dict()` function to take the `strings_by_author` dictionary as an argument, split out the words in the strings, and return a dictionary called `words_by_author` with the authors as keys and a list of words as values ❷.

Stylometry relies on word counts, so it works best when each corpus is the same length. There are multiple ways to ensure apples-to-apples comparisons. With *chunking*, you divide the text into blocks of, say, 5,000 words, and compare the blocks. You can also normalize by using relative frequencies, rather than direct counts, or by truncating to the shortest corpus.

Let's explore the truncation option. Pass the words dictionary to another function, `find_shortest_corpus()`, which calculates the number of words in each author's list and returns the length of the shortest corpus. Table 2-1 shows the length of each corpus.

Table 2-1: Length (Word Count) of Each Corpus

Corpus	Length
Hound (Doyle)	58,387
War (Wells)	59,469
World (Unknown)	74,961

Since the shortest corpus here represents a robust dataset of almost 60,000 words, you'll use the len_shortest_corpus variable to truncate the other two corpora to this length, prior to doing any analysis. The assumption, of course, is that the backend content of the truncated texts is not significantly different from that in the front.

The next five lines call functions that perform the stylometric analysis, as listed in "The Strategy" on page 28 ❸. All the functions take the words_by_author dictionary as an argument, and most take len_shortest_corpus, as well. We'll look at these functions as soon as we finish preparing the texts for analysis.

Loading Text and Building a Word Dictionary

Listing 2-2 defines two functions. The first reads in a text file as a string. The second builds a dictionary with each author's name as the key and his novel, now tokenized into individual words rather than a continuous string, as the value.

```
def text_to_string(filename):
    """Read a text file and return a string."""
    with open(filename) as infile:
        return infile.read()

❶ def make_word_dict(strings_by_author):
    """Return dictionary of tokenized words by corpus by author."""
    words_by_author = dict()
    for author in strings_by_author:
        tokens = nltk.word_tokenize(strings_by_author[author])
    ❷ words_by_author[author] = ([token.lower() for token in tokens
                                    if token.isalpha()])
    return words_by_author
```

Listing 2-2: Defining the text_to_string() and make_word_dict() functions

First, define the text_to_string() function to load a text file. The built-in read() function reads the whole file as an individual string, allowing relatively easy file-wide manipulations. Use with to open the file so that it will be closed automatically regardless of how the block terminates. Just like putting away your toys, closing files is good practice. It prevents bad things from happening, like running out of file descriptors, locking files from further access, corrupting files, or losing data if writing to files.

Some users may encounter a UnicodeDecodeError like the following one when loading the text:

```
UnicodeDecodeError: 'ascii' codec can't decode byte 0x93 in position 365:
ordinal not in range(128)
```

Encoding and *decoding* refer to the process of converting from characters stored as bytes to human-readable strings. The problem is that the default encoding for the built-in function open() is platform dependent and depends on the value of locale.getpreferredencoding(). For example, you'll get the following encoding if you run this on Windows 10:

```
>>> import locale
>>> locale.getpreferredencoding()
'cp1252'
```

CP-1252 is a legacy Windows character encoding. If you run the same code on a Mac, it may return something different, like 'US-ASCII' or 'UTF-8'.

UTF stands for *Unicode Transformational Format*, which is a text character format designed for backward compatibility with ASCII. Although UTF-8 can handle all character sets—and is the dominant form of encoding used on the World Wide Web—it's not the default option for many text editors.

Additionally, Python 2 assumed all text files were encoded with latin-1, used for the Latin alphabet. Python 3 is more sophisticated and tries to detect encoding problems as early as possible. It may throw an error, however, if the encoding isn't specified.

So, the first troubleshooting step should be to pass open() the encoding argument and specify UTF-8.

```
with open(filename, encoding='utf-8') as infile:
```

If you still have problems loading the corpora files, try adding an errors argument as follows:

```
with open(filename, encoding='utf-8', errors='ignore') as infile:
```

You can ignore errors because these text files were downloaded as UTF-8 and have already been tested using this approach. For more on UTF-8, see *https://docs.python.org/3/howto/unicode.html*.

Next, define the make_word_dict() function that will take the dictionary of strings by author and return a dictionary of words by author ❶. First, initialize an empty dictionary named words_by_author. Then, loop through the keys in the strings_by_author dictionary. Use NLTK's word_tokenize() method and pass it the string dictionary's key. The result will be a list of tokens that will serve as the dictionary value for each author. Tokens are just chopped up pieces of a corpus, typically sentences or words.

The following snippet demonstrates how the process turns a continuous string into a list of tokens (words and punctuation):

```
>>> import nltk
>>> str1 = 'The rain in Spain falls mainly on the plain.'
>>> tokens = nltk.word_tokenize(str1)
>>> print(type(tokens))
<class 'list'>
>>> tokens
['The', 'rain', 'in', 'Spain', 'falls', 'mainly', 'on', 'the', 'plain', '.']
```

This is similar to using Python's built-in split() function, but split() doesn't achieve tokens from a linguistic standpoint (note that the period is not tokenized).

```
>>> my_tokens = str1.split()
>>> my_tokens
['The', 'rain', 'in', 'Spain', 'falls', 'mainly', 'on', 'the', 'plain.']
```

Once you have the tokens, populate the words_by_author dictionary using list comprehension ❷. *List comprehension* is a shorthand way to execute loops in Python. You need to surround the code with square brackets to indicate a list. Convert the tokens to lowercase and use the built-in isalpha() method, which returns True if all the characters in a token are part of the alphabet and False otherwise. This will filter out numbers and punctuation. It will also filter out hyphenated words or names. Finish by returning the words_by_author dictionary.

Finding the Shortest Corpus

In computational linguistics, *frequency* refers to the number of occurrences in a corpus. Thus, frequency means the *count*, and methods you'll use later return a dictionary of words and their counts. To compare counts in a meaningful way, the corpora should all have the same number of words.

Because the three corpora used here are large (see Table 2-1), you can safely normalize the corpora by truncating them all to the length of the shortest. Listing 2-3 defines a function that finds the shortest corpus in the words_by_author dictionary and returns its length.

stylometry.py, part 3

```
def find_shortest_corpus(words_by_author):
    """Return length of shortest corpus."""
    word_count = []
    for author in words_by_author:
        word_count.append(len(words_by_author[author]))
        print('\nNumber of words for {} = {}\n'.
            format(author, len(words_by_author[author])))
    len_shortest_corpus = min(word_count)
    print('length shortest corpus = {}\n'.format(len_shortest_corpus))
    return len_shortest_corpus
```

Listing 2-3: Defining the find_shortest_corpus() function

Define the function that takes the `words_by_author` dictionary as an argument. Immediately start an empty list to hold a word count.

Loop through the authors (keys) in the dictionary. Get the length of the value for each key, which is a list object, and append the length to the `word_count` list. The length here represents the number of words in the corpus. For each pass through the loop, print the author's name and the length of his tokenized corpus.

When the loop ends, use the built-in `min()` function to get the lowest count and assign it to the `len_shortest_corpus` variable. Print the answer and then return the variable.

Comparing Word Lengths

Part of an author's distinctive style is the words they use. Faulkner observed that Hemingway never sent a reader running to the dictionary; Hemingway accused Faulkner of using "10-dollar words." Authorial style is expressed in the length of words and in vocabulary, which we'll look at later in the chapter.

Listing 2-4 defines a function to compare the length of words per corpus and plot the results as a frequency distribution. In a frequency distribution, the lengths of words are plotted against the number of counts for each length. For words that are six letters long, for example, one author may have a count of 4,000, and another may have a count of 5,500. A frequency distribution allows comparison across a range of word lengths, rather than just at the average word length.

The function in Listing 2-4 uses list slicing to truncate the word lists to the length of the shortest corpus so the results aren't skewed by the size of the novel.

stylometry.py, part 4

```
def word_length_test(words_by_author, len_shortest_corpus):
    """Plot word length freq by author, truncated to shortest corpus length."""
    by_author_length_freq_dist = dict()
    plt.figure(1)
    plt.ion()

❶   for i, author in enumerate(words_by_author):
        word_lengths = [len(word) for word in words_by_author[author]
                        [:len_shortest_corpus]]
        by_author_length_freq_dist[author] = nltk.FreqDist(word_lengths)
❷       by_author_length_freq_dist[author].plot(15,
                                                 linestyle=LINES[i],
                                                 label=author,
                                                 title='Word Length')
    plt.legend()
    #plt.show()  # Uncomment to see plot while coding.
```

Listing 2-4: Defining the word_length_test() function

All the stylometric functions will use the dictionary of tokens; almost all will use the length of the shortest corpus parameter to ensure consistent sample sizes. Use these variable names as the function parameters.

Start an empty dictionary to hold the frequency distribution of word lengths by author and then start making plots. Since you are going to make multiple plots, start by instantiating a figure object named 1. So that all the plots stay up after creation, turn on the interactive plot mode with plt.ion().

Next, start looping through the authors in the tokenized dictionary ❶. Use the enumerate() function to generate an index for each author that you'll use to choose a line style for the plot. For each author, use list comprehension to get the length of each word in the value list, with the range truncated to the length of the shortest corpus. The result will be a list where each word has been replaced by an integer representing its length.

Now, start populating your new by-author dictionary to hold frequency distributions. Use nltk.FreqDist(), which takes the list of word lengths and creates a data object of word frequency information that can be plotted.

You can plot the dictionary directly using the class method plot(), without the need to reference pyplot through plt ❷. This will plot the most frequently occurring sample first, followed by the number of samples you specify, in this case, 15. This means you will see the frequency distribution of words from 1 to 15 letters long. Use i to select from the LINES list and finish by providing a label and a title. The label will be used in the legend, called using plt.legend().

Note that you can change how the frequency data plots using the cumulative parameter. If you specify cumulative=True, you will see a cumulative distribution (Figure 2-3, left). Otherwise, plot() will default to cumulative=False, and you will see the actual counts, arranged from highest to lowest (Figure 2-3, right). Continue to use the default option for this project.

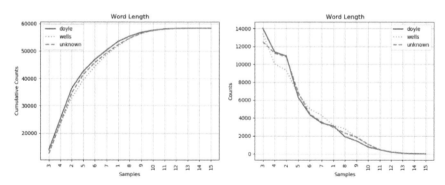

Figure 2-3: The NLTK cumulative plot (left) versus the default frequency plot (right)

Finish by calling the plt.show() method to display the plot, but leave it commented out. If you want to see the plot immediately after coding this function, you can uncomment it. Also note that if you launch this program via Windows PowerShell, the plots may close immediately unless you use the block flag: `plt.show(block=True)`. This will keep the plot up but halt execution of the program until the plot is closed.

Based solely on the word length frequency plot in Figure 2-3, Doyle's style matches the unknown author's more closely, though there are segments where Wells compares the same or better. Now let's run some other tests to see whether we can confirm that finding.

Comparing Stop Words

A *stop word* is a small word used often, like *the*, *by*, and *but*. These words are filtered out for tasks like online searches, because they provide no contextual information, and they were once thought to be of little value in identifying authorship.

But stop words, used frequently and without much thought, are perhaps the best signature for an author's style. And since the texts you're comparing are usually about different subjects, these stop words become important, as they are agnostic to content and common across all texts.

Listing 2-5 defines a function to compare the use of stop words in the three corpora.

stylometry.py, part 5

```
def stopwords_test(words_by_author, len_shortest_corpus):
    """Plot stopwords freq by author, truncated to shortest corpus length."""
    stopwords_by_author_freq_dist = dict()
    plt.figure(2)
    stop_words = set(stopwords.words('english'))  # Use set for speed.
    #print('Number of stopwords = {}\n'.format(len(stop_words)))
    #print('Stopwords = {}\n'.format(stop_words))

    for i, author in enumerate(words_by_author):
        stopwords_by_author = [word for word in words_by_author[author]
                               [:len_shortest_corpus] if word in stop_words]
        stopwords_by_author_freq_dist[author] = nltk.FreqDist(stopwords_by_
        author)
        stopwords_by_author_freq_dist[author].plot(50,
                                                   label=author,
                                                   linestyle=LINES[i],
                                                   title=
                                                   '50 Most Common Stopwords')
    plt.legend()
##    plt.show()  # Uncomment to see plot while coding function.
```

Listing 2-5: Defining the `stopwords_test()` function

Define a function that takes the words dictionary and the length of the shortest corpus variables as arguments. Then initialize a dictionary to hold the frequency distribution of stop words for each author. You don't want to cram all the plots in the same figure, so start a new figure named 2.

Assign a local variable, stop_words, to the NLTK stop words corpus for English. Sets are quicker to search than lists, so make the corpus a set for faster lookups later. The next two lines, currently commented out, print the number of stop words (179) and the stop words themselves.

Now, start looping through the authors in the words_by_author dictionary. Use list comprehension to pull out all the stop words in each author's corpus and use these as the value in a new dictionary named stopwords_by_author. In the next line, you'll pass this dictionary to NLTK's FreqDist() method and use the output to populate the stopwords_by_author_freq_dist dictionary. This dictionary will contain the data needed to make the frequency distribution plots for each author.

Repeat the code you used to plot the word lengths in Listing 2-4, but set the number of samples to 50 and give it a different title. This will plot the top 50 stop words in use (Figure 2-4).

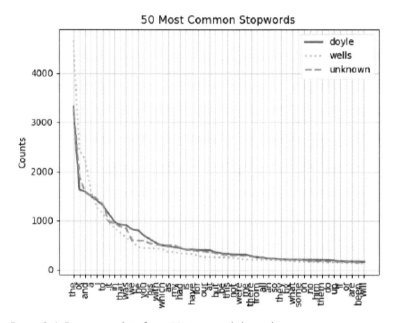

Figure 2-4: Frequency plot of top 50 stop words by author

Both Doyle and the unknown author use stop words in a similar manner. At this point, two analyses have favored Doyle as the most likely author of the unknown text, but there's still more to do.

Comparing Parts of Speech

Now let's compare the parts of speech used in the three corpora. NLTK uses a part-of-speech (POS) tagger, called PerceptronTagger, to identify parts of speech. POS taggers process a sequence of tokenized words and attach a POS tag to each word (see Table 2-2).

Table 2-2: Parts of Speech with Tag Values

Part of Speech	Tag	Part of Speech	Tag
Coordinating conjunction	CC	Possessive pronoun	PRP$
Cardinal number	CD	Adverb	RB
Determiner	DT	Adverb, comparative	RBR
Existential there	EX	Adverb, superlative	RBS
Foreign word	FW	Particle	RP
Preposition or subordinating conjunction	IN	Symbol	SYM
Adjective	JJ	To	TO
Adjective, comparative	JJR	Interjection	UH
Adjective, superlative	JJS	Verb, base form	VB
List item marker	LS	Verb, past tense	VBD
Modal	MD	Verb, gerund or present participle	VBG
Noun, singular or mass	NN	Verb, past participle	VBN
Noun, plural	NNS	Verb, non-third-person singular present	VBP
Noun, proper noun, singular	NNP	Verb, third-person singular present	VBZ
Noun, proper noun, plural	NNPS	Wh-determiner, which	WDT
Predeterminer	PDT	Wh-pronoun, who, what	WP
Possessive ending	POS	Possessive wh-pronoun, whose	WP$
Personal pronoun	PRP	Wh-adverb, where, when	WRB

The taggers are typically trained on large datasets like the *Penn Treebank* or *Brown Corpus*, making them highly accurate though not perfect. You can also find training data and taggers for languages other than English. You don't need to worry about all these various terms and their abbreviations. As with the previous tests, you'll just need to compare lines in a chart.

Listing 2-6 defines a function to plot the frequency distribution of POS in the three corpora.

stylometry.py, part 6

```python
def parts_of_speech_test(words_by_author, len_shortest_corpus):
    """Plot author use of parts-of-speech such as nouns, verbs, adverbs."""
    by_author_pos_freq_dist = dict()
    plt.figure(3)
    for i, author in enumerate(words_by_author):
        pos_by_author = [pos[1] for pos in nltk.pos_tag(words_by_author[author]
                         [:len_shortest_corpus])]
        by_author_pos_freq_dist[author] = nltk.FreqDist(pos_by_author)
        by_author_pos_freq_dist[author].plot(35,
                                             label=author,
                                             linestyle=LINES[i],
                                             title='Part of Speech')
    plt.legend()
    plt.show()
```

Listing 2-6: Defining the parts_of_speech_test() function

Define a function that takes as arguments—you guessed it—the words dictionary and the length of the shortest corpus. Then initialize a dictionary to hold the frequency distribution for the POS for each author, followed by a function call for a third figure.

Start looping through the authors in the words_by_author dictionary and use list comprehension and the NLTK pos_tag() method to build a list called pos_by_author. For each author, this creates a list with each word in the author's corpus replaced by its corresponding POS tag, as shown here:

```
['NN', 'NNS', 'WP', 'VBD', 'RB', 'RB', 'RB', 'IN', 'DT', 'NNS', --snip--]
```

Next, make a frequency distribution of the POS list and with each loop plot the curve, using the top 35 samples. Note that there are only 36 POS tags and several, such as *list item markers*, rarely appear in novels.

This is the final plot you'll make, so call plt.show() to draw all the plots to the screen. As pointed out in the discussion of Listing 2-4, if you're using Windows PowerShell to launch the program, you may need to use plt.show(block=True) to keep the plots from closing automatically.

The previous plots, along with the current one (Figure 2-5), should appear after about 10 seconds.

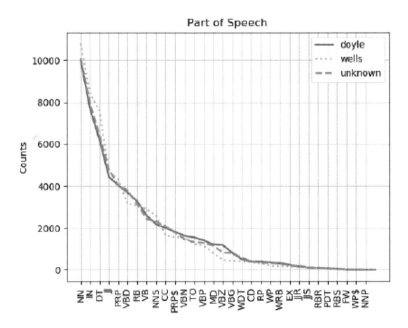

Figure 2-5: Frequency plot of top 35 parts of speech by author

Once again, the match between the Doyle and unknown curves is clearly better than the match of unknown to Wells. This suggests that Doyle is the author of the unknown corpus.

Comparing Author Vocabularies

To compare the vocabularies among the three corpora, you'll use the *chi-squared random variable* (X^2), also known as the *test statistic*, to measure the "distance" between the vocabularies employed in the unknown corpus and each of the known corpora. The closest vocabularies will be the most similar. The formula is

$$X^2 = \sum_{i=1}^{n} \frac{(O_i - E_i)^2}{E_i}$$

where *O* is the observed word count and *E* is the expected word count assuming the corpora being compared are both by the same author.

If Doyle wrote both novels, they should both have the same—or a similar—proportion of the most common words. The test statistic lets you quantify how similar they are by measuring how much the counts for each word differ. The lower the chi-squared test statistic, the greater the similarity between two distributions.

Listing 2-7 defines a function to compare vocabularies among the three corpora.

stylometry.py, part 7

```
def vocab_test(words_by_author):
    """Compare author vocabularies using the chi-squared statistical test."""
    chisquared_by_author = dict()
    for author in words_by_author:
❶      if author != 'unknown':
            combined_corpus = (words_by_author[author] +
                               words_by_author['unknown'])
            author_proportion = (len(words_by_author[author])/
                                 len(combined_corpus))
            combined_freq_dist = nltk.FreqDist(combined_corpus)
            most_common_words = list(combined_freq_dist.most_common(1000))
            chisquared = 0
❷          for word, combined_count in most_common_words:
                observed_count_author = words_by_author[author].count(word)
                expected_count_author = combined_count * author_proportion
                chisquared += ((observed_count_author -
                               expected_count_author)**2 /
                               expected_count_author)
❸          chisquared_by_author[author] = chisquared
            print('Chi-squared for {} = {:.1f}'.format(author, chisquared))
    most_likely_author = min(chisquared_by_author, key=chisquared_by_author.get)
    print('Most-likely author by vocabulary is {}\n'.format(most_likely_author))
```

Listing 2-7: Defining the vocab_test() function

The vocab_test() function needs the word dictionary but not the length of the shortest corpus. Like the previous functions, however, it starts by creating a new dictionary to hold the chi-squared value per author and then loops through the word dictionary.

To calculate chi-squared, you'll need to join each author's corpus with the unknown corpus. You don't want to combine unknown with itself, so use a conditional to avoid this ❶. For the current loop, combine the author's corpus with the unknown one and then get the current author's proportion by dividing the length of his corpus by the length of the combined corpus. Then get the frequency distribution of the combined corpus by calling nltk.FreqDist().

Now, make a list of the 1,000 most common words in the combined text by using the most_common() method and passing it 1000. There is no hard-and-fast rule for how many words you should consider in a stylometric analysis. Suggestions in the literature call for the most common 100 to 1,000 words. Since you are working with large texts, err on the side of the larger value.

Initialize the chisquared variable with 0; then start a nested for loop that works through the most_common_words list ❷. The most_common() method returns a list of tuples, with each tuple containing the word and its count.

```
[('the', 7778), ('of', 4112), ('and', 3713), ('i', 3203), ('a', 3195), --snip--]
```

Next, you get the observed count per author from the word dictionary. For Doyle, this would be the count of the most common words in the corpus of *The Hound of the Baskervilles*. Then, you get the expected count, which for Doyle would be the count you would expect if he wrote both *The Hound of the Baskervilles* and the unknown corpus. To do this, multiply the number of counts in the combined corpus by the previously calculated author's proportion. Then apply the formula for chi-squared and add the result to the dictionary that tracks each author's chi-squared score ❸. Display the result for each author.

To find the author with the lowest chi-squared score, call the built-in min() function and pass it the dictionary and dictionary key, which you obtain with the get() method. This will yield the *key* corresponding to the minimum *value*. This is important. If you omit this last argument, min() will return the minimum *key* based on the alphabetical order of the names, *not* their chi-squared score! You can see this mistake in the following snippet:

```
>>> print(mydict)
{'doyle': 100, 'wells': 5}
>>> minimum = min(mydict)
>>> print(minimum)
'doyle'
>>> minimum = min(mydict, key=mydict.get)
>>> print(minimum)
'wells'
```

It's easy to assume that the min() function returns the minimum numerical *value*, but as you saw, it looks at dictionary *keys* by default.

Complete the function by printing the most likely author based on the chi-squared score.

```
Chi-squared for doyle = 4744.4
Chi-squared for wells = 6856.3
Most-likely author by vocabulary is doyle
```

Yet another test suggests that Doyle is the author!

Calculating Jaccard Similarity

To determine the degree of similarity among sets created from the corpora, you'll use the *Jaccard similarity coefficient*. Also called the *intersection over union*, this is simply the area of overlap between two sets divided by the area of union of the two sets (Figure 2-6).

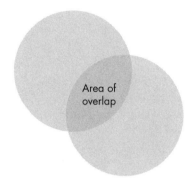

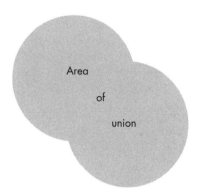

Figure 2-6: Intersection-over-union for a set is the area of overlap divided by the area of union.

The more overlap there is between sets created from two texts, the more likely they were written by the same author. Listing 2-8 defines a function for gauging the similarity of sample sets.

stylometry.py, part 8

```
def jaccard_test(words_by_author, len_shortest_corpus):
    """Calculate Jaccard similarity of each known corpus to unknown corpus."""
    jaccard_by_author = dict()
    unique_words_unknown = set(words_by_author['unknown']
                               [:len_shortest_corpus])
```

```
❶ authors = (author for author in words_by_author if author != 'unknown')
  for author in authors:
      unique_words_author = set(words_by_author[author][:len_shortest_corpus])
      shared_words = unique_words_author.intersection(unique_words_unknown)
❷ jaccard_sim = (float(len(shared_words))/ (len(unique_words_author) +
                                            len(unique_words_unknown) -
                                            len(shared_words)))
      jaccard_by_author[author] = jaccard_sim
      print('Jaccard Similarity for {} = {}'.format(author, jaccard_sim))
❸ most_likely_author = max(jaccard_by_author, key=jaccard_by_author.get)
  print('Most-likely author by similarity is {}'.format(most_likely_author))

if __name__ == '__main__':
    main()
```

Listing 2-8: Defining the jaccard_test() function

Like most of the previous tests, the jaccard_test() function takes the word dictionary and length of the shortest corpus as arguments. You'll also need a dictionary to hold the Jaccard coefficient for each author.

Jaccard similarity works with unique words, so you'll need to turn the corpora into sets to remove duplicates. First, you'll build a set from the unknown corpus. Then you'll loop through the known corpora, turning them into sets and comparing them to the unknown set. Be sure to truncate all the corpora to the length of the shortest corpus when making the sets.

Prior to running the loop, use a generator expression to get the names of the authors, other than unknown, from the words_by_author dictionary ❶. A *generator expression* is a function that returns an object that you can iterate over one value at a time. It looks a lot like list comprehension, but instead of square brackets, it's surrounded by parentheses. And instead of constructing a potentially memory-intensive list of items, the generator yields them in real time. Generators are useful when you have a large set of values that you need to use only once. I use one here as an opportunity to demonstrate the process.

When you assign a generator expression to a variable, all you get is a type of iterator called a *generator object*. Compare this to making a list, as shown here:

```
>>> mylist = [i for i in range(4)]
>>> mylist
[0, 1, 2, 3]
>>> mygen = (i for i in range(4))
>>> mygen
<generator object <genexpr> at 0x000002717F547390>
```

The generator expression in the previous snippet is the same as this generator function:

```
def generator(my_range):
    for i in range(my_range):
        yield i
```

Whereas the return statement ends a function, the yield statement *suspends* the function's execution and sends a value back to the caller. Later, the function can resume where it left off. When a generator reaches its end, it's "empty" and can't be called again.

Back to the code, start a for loop using the authors generator. Find the unique words for each known author, just as you did for unknown. Then use the built-in intersection() function to find all the words shared between the current author's set of words and the set for unknown. The *intersection* of two given sets is the largest set that contains all the elements that are common to both. With this information, you can calculate the Jaccard similarity coefficient ❷.

Update the jaccard_by_author dictionary and print each outcome in the interpreter window. Then find the author with the maximum Jaccard value ❸ and print the results.

```
Jaccard Similarity for doyle = 0.34847801578354004
Jaccard Similarity for wells = 0.30786921307869214
Most-likely author by similarity is doyle
```

The outcome should favor Doyle.

Finish *stylometry.py* with the code to run the program as an imported module or in stand-alone mode.

Summary

The true author of *The Lost World* is Doyle, so we'll stop here and declare victory. If you want to explore further, a next step might be to add more known texts to doyle and wells so that their combined length is closer to that for *The Lost World* and you don't have to truncate it. You could also test for sentence length and punctuation style or employ more sophisticated techniques like neural nets and genetic algorithms.

You can also refine existing functions, like vocab_test() and jaccard_test(), with *stemming* and *lemmatization* techniques that reduce words to their root forms for better comparisons. As the program is currently written, *talk*, *talking*, and *talked* are all considered completely different words even though they share the same root.

At the end of the day, stylometry can't prove with absolute certainty that Sir Arthur Conan Doyle wrote *The Lost World*. It can only suggest, through weight of evidence, that he is the more likely author than Wells. Framing the question very specifically is important, since you can't evaluate all possible authors. For this reason, successful authorship attribution begins with good old-fashioned detective work that trims the list of candidates to a manageable length.

Further Reading

Natural Language Processing with Python: Analyzing Text with the Natural Language Toolkit (O'Reilly, 2009), by Steven Bird, Ewan Klein, and Edward Loper, is an accessible introduction to NLP using Python, with lots of exercises and useful integration with the NLTK website. A new version of the book, updated for Python 3 and NLTK 3, is available online at *http://www.nltk.org/book/*.

In 1995, novelist Kurt Vonnegut proposed the idea that "stories have shapes that can be drawn on graph paper" and suggested "feeding them into computers." In 2018, researchers followed up on this idea using more than 1,700 English novels. They applied an NLP technique called *sentiment analysis* that finds the emotional tone behind words. An interesting summary of their results, "Every Story in the World Has One of These Six Basic Plots," can be found on the BBC.com website: *http://www.bbc.com/culture/story/20180525-every-story-in-the-world-has-one-of-these-six-basic-plots/*.

Practice Project: Hunting the Hound with Dispersion

NLTK comes with a fun little feature, called a *dispersion plot*, that lets you post the location of a word in a text. More specifically, it plots the occurrences of a word versus how many words from the beginning of the corpus that it appears.

Figure 2-7 is a dispersion plot for major characters in *The Hound of the Baskervilles*.

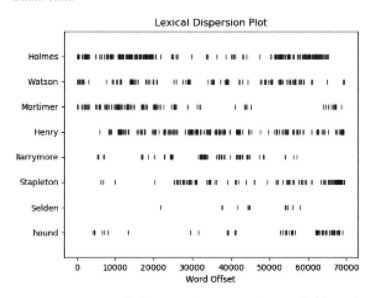

Figure 2-7: Dispersion plot for major characters in The Hound of the Baskervilles

If you're familiar with the story—and I won't spoil it if you're not—then you'll appreciate the sparse occurrence of Holmes in the middle, the almost bimodal distribution of Mortimer, and the late story overlap of Barrymore, Selden, and the hound.

Dispersion plots can have more practical applications. For example, as the author of technical books, I need to define a new term when it first appears. This sounds easy, but sometimes the editing process can shuffle whole chapters, and issues like this can fall through the cracks. A dispersion plot, built with a long list of technical terms, can make finding these first occurrences a lot easier.

For another use case, imagine you're a data scientist working with para-legals on a criminal case involving insider trading. To find out whether the accused talked to a certain board member just prior to making the illegal trades, you can load the subpoenaed emails of the accused as a continuous string and generate a dispersion plot. If the board member's name appears as expected, case closed!

For this practice project, write a Python program that reproduces the dispersion plot shown in Figure 2-7. If you have problems loading the *hound.txt* corpus, revisit the discussion of Unicode on page 35. You can find a solution, *practice_hound_dispersion.py*, in the appendix and online.

Practice Project: Punctuation Heatmap

A *heatmap* is a diagram that uses colors to represent data values. Heatmaps have been used to visualize the punctuation habits of famous authors (*https://www.fastcompany.com/3057101/the-surprising-punctuation-habits-of-famous-authors-visualized/*) and may prove helpful in attributing authorship for *The Lost World*.

Write a Python program that tokenizes the three novels used in this chapter based solely on punctuation. Then focus on the use of semicolons. For each author, plot a heatmap that displays semicolons as blue and all other marks as yellow or red. Figure 2-8 shows example heatmaps for Wells' *The War of the Worlds* and Doyle's *The Hound of the Baskervilles*.

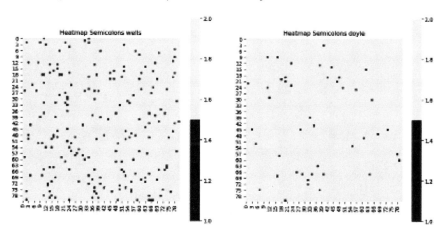

Figure 2-8: Heatmap of semicolon use (dark squares) for Wells (left) and Doyle (right)

Compare the three heatmaps. Do the results favor Doyle or Wells as the author for *The Lost World*?

You can find a solution, *practice_heatmap_semicolon.py*, in the appendix and online.

Challenge Project: Fixing Frequency

As noted previously, frequency in NLP refers to counts, but it can also be expressed as the number of occurrences per unit time. Alternatively, it can be expressed as a ratio or percent.

Define a new version of the nltk.FreqDist() method that uses percentages, rather than counts, and use it to make the charts in the *stylometry.py* program. For help, see the Clearly Erroneous blog (*https://martinapugliese .github.io/plotting-the-actual-frequencies-in-a-FreqDist-in-nltk/*).

3

SUMMARIZING SPEECHES WITH NATURAL LANGUAGE PROCESSING

"Water, water everywhere, but not a drop to drink." This famous line, from *The Rime of the Ancient Mariner,* summarizes the present state of digital information. According to International Data Corporation, by 2025 we'll be generating 175 trillion gigabytes of digital data *per year.* But most of this data—up to 95 percent—will be *unstructured,* which means it's not organized into useful databases. Even now, the key to the cure for cancer may be right at our fingertips yet almost impossible to reach.

To make information easier to discover and consume, we need to reduce the volume of data by extracting and repackaging salient points into digestible summaries. Because of the sheer volume of data, there's no way to do this manually. Luckily, natural language processing (NLP) helps computers understand both words and context. For example, NLP applications can summarize news feeds, analyze legal contracts, research patents, study financial markets, capture corporate knowledge, and produce study guides.

In this chapter, you'll use Python's Natural Language Toolkit (NLTK) to generate a summary of one of the most famous speeches of all time, "I Have a Dream" by Martin Luther King Jr. With an understanding of the basics, you'll then use a streamlined alternative, called gensim, to summarize the popular "Make Your Bed" speech by Admiral William H. McRaven. Finally, you'll use a word cloud to produce a fun visual summary of the most frequently used words in Sir Arthur Conan Doyle's novel *The Hound of the Baskervilles*.

Project #3: I Have a Dream . . . to Summarize Speeches!

In machine learning and data mining, there are two approaches to summarizing text: *extraction* and *abstraction*.

Extraction-based summarization uses various weighting functions to rank sentences by perceived importance. Words used more often are considered more important. Consequently, sentences containing those words are considered more important. The overall behavior is like using a yellow highlighter to manually select keywords and sentences without altering the text. The results can be disjointed, but the technique is good at pulling out important words and phrases.

Abstraction relies on deeper comprehension of the document to capture intent and produce more human-like paraphrasing. This includes creating completely new sentences. The results tend to be more cohesive and grammatically correct than those produced by extraction-based methods, but at a price. Abstraction algorithms require advanced and complicated deep learning methods and sophisticated language modeling.

For this project, you'll use an extraction-based technique on the "I Have a Dream" speech, delivered by Martin Luther King Jr. at the Lincoln Memorial on August 28, 1963. Like Lincoln's "Gettysburg Address" a century before, it was the perfect speech at the perfect time. Dr. King's masterful use of repetition also makes it tailor-made for extraction techniques, which correlate word frequency with importance.

THE OBJECTIVE

Write a Python program that summarizes a speech using NLP text extraction.

The Strategy

The Natural Language Toolkit includes the functions you'll need to summarize Dr. King's speech. If you skipped Chapter 2, see page 29 for installation instructions.

To summarize the speech, you'll need a digital copy. In previous chapters, you manually downloaded files you needed from the internet.

This time you'll use a more efficient technique, called *web scraping*, which allows you to programmatically extract and save large amounts of data from websites.

Once you've loaded the speech as a string, you can use NLTK to split out and count individual words. Then, you'll "score" each sentence in the speech by summing the word counts within it. You can use those scores to print the top-ranked sentences, based on how many sentences you want in your summary.

Web Scraping

Scraping the web means using a program to download and process content. This is such a common task that prewritten scraping programs are freely available. You'll use the requests library to download files and web pages, and you'll use the Beautiful Soup (bs4) package to parse HTML. Short for *Hypertext Markup Language*, HTML is the standard format used to create web pages.

To install the two modules, use pip in a terminal window or Windows PowerShell (see page 8 in Chapter 1 for instructions on installing and using pip):

```
pip install requests
pip install beautifulsoup4
```

To check the installation, open the shell and import each module as shown next. If you don't get an error, you're good to go!

```
>>> import requests
>>>
>>> import bs4
>>>
```

To learn more about requests, visit *https://pypi.org/project/requests/*. For Beautiful Soup, see *https://www.crummy.com/software/BeautifulSoup/*.

The "I Have a Dream" Code

The *dream_summary.py* program performs the following steps:

1. Opens a web page containing the "I Have a Dream" speech
2. Loads the text as a string
3. Tokenizes the text into words and sentences
4. Removes stop words with no contextual content
5. Counts the remaining words
6. Uses the counts to rank the sentences
7. Displays the highest-ranking sentences

If you've already downloaded the book's files, find the program in the *Chapter_3* folder. Otherwise, go to *https://nostarch.com/real-world-python/* and download it from the book's GitHub page.

Importing Modules and Defining the main() Function

Listing 3-1 imports modules and defines the first part of the main() function, which scrapes the web and assigns the speech to a variable as a string.

*dream_*
summary.py,
part 1

```
from collections import Counter
import re
import requests
import bs4
import nltk
from nltk.corpus import stopwords

def main():
❶ url = 'http://www.analytictech.com/mb021/mlk.htm'
  page = requests.get(url)
  page.raise_for_status()
❷ soup = bs4.BeautifulSoup(page.text, 'html.parser')
  p_elems = [element.text for element in soup.find_all('p')]

  speech = ''.join(p_elems)
```

Listing 3-1: Importing modules and defining the main() function

Start by importing Counter from the collections module to help you keep track of the sentence scoring. The collections module is part of the Python Standard Library and includes several container data types. A Counter is a dictionary subclass for counting hashable objects. Elements are stored as dictionary keys, and their counts are stored as dictionary values.

Next, to clean up the speech prior to summarizing its contents, import the re module. The *re* stands for *regular expressions*, also referred to as *regexes*, which are sequences of characters that define a search pattern. This module will help you clean up the speech by allowing you to selectively remove bits that you don't want.

Finish the imports with the modules for scraping the web and doing natural language processing. The last module brings in the list of functional stop words (such as *if, and, but, for*) that contain no useful information. You'll remove these from the speech prior to summarization.

Next, define a main() function to run the program. To scrape the speech off the web, provide the url address as a string ❶. You can copy and paste this from the website from which you want to extract text.

The requests library abstracts the complexities of making HTTP requests in Python. HTTP, short for HyperText Transfer Protocol, is the foundation of data communication using hyperlinks on the World Wide Web. Use the requests.get() method to fetch the url and assign the output to the page variable, which references the Response object the web page returned for the request. This object's text attribute holds the web page, including the speech, as a string.

To check that the download was successful, call the Response object's raise_for_status() method. This does nothing if everything goes okay but otherwise will raise an exception and halt the program.

At this point, the data is in HTML, as shown here:

```
<!DOCTYPE HTML PUBLIC "-//IETF//DTD HTML//EN">
<html>

<head>
<meta http-equiv="Content-Type"
content="text/html; charset=iso-8859-1">
<meta name="GENERATOR" content="Microsoft FrontPage 4.0">
<title>Martin Luther King Jr.'s 1962 Speech</title>
</head>
--snip--
<p>I am happy to join with you today in what will go down in
history as the greatest demonstration for freedom in the history
of our nation. </p>
--snip--
```

As you can see, HTML has a lot of *tags*, such as <head> and <p>, that let your browser know how to format the web page. The text between starting and closing tags is called an *element*. For example, the text "Martin Luther King Jr.'s 1962 Speech" is a title element sandwiched between the starting tag <title> and the closing tag </title>. Paragraphs are formatted using <p> and </p> tags.

Because these tags are not part of the original text, they should be removed prior to any natural language processing. To remove the tags, call the bs4.BeautifulSoup() method and pass it the string containing the HTML ❷. Note that I've explicitly specified html.parser. The program will run without this but complain bitterly with warnings in the shell.

The soup variable now references a BeautifulSoup object, which means you can use the object's find_all() method to locate the speech buried in the HTML document. In this case, to find the text between paragraph tags (<p>), use list comprehension and find_all() to make a list of just the paragraph elements.

Finish by turning the speech into a continuous string. Use the join() method to turn the p_elems list into a string. Set the "joiner" character to a space, designated by ' '.

Note that with Python, there is usually more than one way to accomplish a task. The last two lines of the listing can also be written as follows:

```
p_elems = soup.select('p')
speech = ''.join(p_elems)
```

The select() method is more limited overall than find_all(), but in this case it works the same and is more succinct. In the previous snippet, select() finds the <p> tags, and the results are converted to text when concatenated to the speech string.

Completing the main() Function

Next, you'll prep the speech to fix typos and remove punctuation, special characters, and spaces. Then you'll call three functions to remove stop words, count word frequency, and score the sentences based on the word counts. Finally, you'll rank the sentences and display those with the highest scores in the shell.

Listing 3-2 completes the definition of main() that performs these tasks.

*dream_*
summary.py,
part 2

```
speech = speech.replace(')mowing', 'knowing')
speech = re.sub('\s+', ' ', speech)
speech_edit = re.sub('[^a-zA-Z]', ' ', speech)
speech_edit = re.sub('\s+', ' ', speech_edit)

❶ while True:
      max_words = input("Enter max words per sentence for summary: ")
      num_sents = input("Enter number of sentences for summary: ")
      if max_words.isdigit() and num_sents.isdigit():
          break
      else:
          print("\nInput must be in whole numbers.\n")

  speech_edit_no_stop = remove_stop_words(speech_edit)
  word_freq = get_word_freq(speech_edit_no_stop)
  sent_scores = score_sentences(speech, word_freq, max_words)

❷ counts = Counter(sent_scores)
  summary = counts.most_common(int(num_sents))
  print("\nSUMMARY:")
  for i in summary:
      print(i[0])
```

Listing 3-2: Completing the main() function

The original document contains a typo (*mowing* instead of *knowing*), so start by fixing this using the string.replace() method. Continue cleaning the speech using regex. Many casual programmers are turned off by this module's arcane syntax, but it's such a powerful and useful tool that everyone should be aware of the basic regex syntax.

Remove extra spaces using the re.sub() function, which replaces substrings with new characters. Use the shorthand character class code \s+ to identify runs of whitespace and replace them with a single space, indicated by ' '. Finish by passing re.sub() the name of the string (speech).

Next, remove anything that's *not* a letter by matching the [^a-zA-Z] pattern. The caret at the start instructs regex to "match any character that isn't between the brackets." So, numbers, punctuation marks, and so on, will be replaced by a space.

Removing characters like punctuation marks will leave an extra space. To get rid of these spaces, call the re.sub() method again.

Next, request that the user input the number of sentences to include in the summary and the maximum number of words per sentence. Use a while loop and Python's built-in isdigit() function to ensure the user inputs an integer ❶.

NOTE *According to research by the American Press Institute, comprehension is best with sentences of fewer than 15 words. Similarly, the* Oxford Guide to Plain English *recommends using sentences that average 15 to 20 words over a full document.*

Continue cleaning the text by calling the `remove_stop_words()` function. Then call functions `get_word_freq()` and `score_sentences()` to calculate the frequency of the remaining words and to score the sentences, respectively. You'll define these functions after completing the `main()` function.

To rank the sentences, call the `collection` module's `Counter()` method ❷. Pass it the `sent_scores` variable.

To generate the summary, use the `Counter` object's `most_common()` method. Pass it the `num_sents` variable input by the user. The resulting `summary` variable will hold a list of tuples. For each tuple, the sentence is at index `[0]`, and its rank is at index `[1]`.

```
[('From every mountainside, let freedom ring.', 4.625), --snip-- ]
```

For readability, print each sentence of the summary on a separate line.

Removing Stop Words

Remember from Chapter 2 that stop words are short, functional words like *if, but, for,* and *so.* Because they contain no important contextual information, you don't want to use them to rank sentences.

Listing 3-3 defines a function called `remove_stop_words()` to remove stop words from the speech.

*dream_
summary.py,
part 3*

```
def remove_stop_words(speech_edit):
    """Remove stop words from string and return string."""
    stop_words = set(stopwords.words('english'))
    speech_edit_no_stop = ''
    for word in nltk.word_tokenize(speech_edit):
        if word.lower() not in stop_words:
            speech_edit_no_stop += word + ' '
    return speech_edit_no_stop
```

Listing 3-3: Defining a function to remove stop words from the speech

Define the function to receive `speech_edit`, the edited speech string, as an argument. Then create a set of the English stop words in NLTK. Use a set, rather than a list, as searches are quicker in sets.

Assign an empty string to hold the edited speech sans stop words. The `speech_edit` variable is currently a string in which each element is a letter.

To work with words, call the NLTK `word_tokenize()` method. Note that you can do this while looping through words. Convert each word to lowercase and check its membership in the `stop_words` set. If it's not a stop word, concatenate it to the new string, along with a space. Return this string to end the function.

How you handle letter case in this program is important. You'll want the summary to print with both uppercase and lowercase letters, but you must do the NLP work using all lowercase to avoid miscounting. To see why, look at the following code snippet, which counts words in a string (s) with mixed cases:

```
>>> import nltk
>>> s = 'one One one'
>>> fd = nltk.FreqDist(nltk.word_tokenize(s))
>>> fd
FreqDist({'one': 2, 'One': 1})
>>> fd_lower = nltk.FreqDist(nltk.word_tokenize(s.lower()))
>>> fd_lower
FreqDist({'one': 3})
```

If you don't convert the words to lowercase, *one* and *One* are considered distinct elements. For counting purposes, every instance of *one* regardless of its case should be treated as the same word. Otherwise, the contribution of *one* to the document will be diluted.

Calculating the Frequency of Occurrence of Words

To count the occurrence of each word in the speech, you'll create the get_word_freq() function that returns a dictionary with the words as keys and the counts as values. Listing 3-4 defines this function.

*dream_
summary.py,
part 4*

```
def get_word_freq(speech_edit_no_stop):
    """Return a dictionary of word frequency in a string."""
    word_freq = nltk.FreqDist(nltk.word_tokenize(speech_edit_no_stop.lower()))
    return word_freq
```

Listing 3-4: Defining a function to calculate word frequency in the speech

The get_word_freq() function takes the edited speech string with no stop words as an argument. NLTK's FreqDist class acts like a dictionary with the words as keys and their counts as values. As part of the process, convert the input string to lowercase and tokenize it into words. End the function by returning the word_freq dictionary.

Scoring Sentences

Listing 3-5 defines a function that scores sentences based on the frequency distribution of the words they contain. It returns a dictionary with each sentence as the key and its score as the value.

*dream_
summary.py,
part 5*

```
def score_sentences(speech, word_freq, max_words):
    """Return dictionary of sentence scores based on word frequency."""
    sent_scores = dict()
    sentences = nltk.sent_tokenize(speech)
❶   for sent in sentences:
        sent_scores[sent] = 0
```

```
        words = nltk.word_tokenize(sent.lower())
        sent_word_count = len(words)
❷      if sent_word_count <= int(max_words):
            for word in words:
                if word in word_freq.keys():
                    sent_scores[sent] += word_freq[word]
❸          sent_scores[sent] = sent_scores[sent] / sent_word_count
    return sent_scores

if __name__ == '__main__':
    main()
```

Listing 3-5: Defining a function to score sentences based on word frequency

Define a function, called score_sentences(), with parameters for the original speech string, the word_freq object, and the max_words variable input by the user. You want the summary to contain stop words and capitalized words—hence the use of speech.

Start an empty dictionary, named sent_scores, to hold the scores for each sentence. Next, tokenize the speech string into sentences.

Now, start looping through the sentences ❶. Start by updating the sent_scores dictionary, assigning the sentence as the key, and setting its initial value (count) to 0.

To count word frequency, you first need to tokenize the sentence into words. Be sure to use lowercase to be compatible with the word_freq dictionary.

You'll need to be careful when you sum up the word counts per sentence to create the scores so you don't bias the results toward longer sentences. After all, longer sentences are more likely to have a greater number of important words. To avoid excluding short but important sentences, you need to *normalize* each count by dividing it by the sentence *length*. Store the length in a variable called sent_word_count.

Next, use a conditional that constrains sentences to the maximum length input by the user ❷. If the sentence passes the test, start looping through its words. If a word is in the word_freq dictionary, add it to the count stored in sent_scores.

At the end of each loop through the sentences, divide the score for the current sentence by the number of words in the sentence ❸. This normalizes the score so long sentences don't have an unfair advantage.

End the function by returning the sent_scores dictionary. Then, back in the global space, add the code for running the program as a module or in stand-alone mode.

Running the Program

Run the *dream_summary.py* program with a maximum sentence length of 14 words. As mentioned previously, good, readable sentences tend to contain 14 words or fewer. Then truncate the summary at 15 sentences, about one-third of the speech. You should get the following results. Note that the sentences won't necessarily appear in their original order.

```
Enter max words per sentence for summary: 14
Enter number of sentences for summary: 15

SUMMARY:
From every mountainside, let freedom ring.
Let freedom ring from Lookout Mountain in Tennessee!
Let freedom ring from every hill and molehill in Mississippi.
Let freedom ring from the curvaceous slopes of California!
Let freedom ring from the snow capped Rockies of Colorado!
But one hundred years later the Negro is still not free.
From the mighty mountains of New York, let freedom ring.
From the prodigious hilltops of New Hampshire, let freedom ring.
And I say to you today my friends, let freedom ring.
I have a dream today.
It is a dream deeply rooted in the American dream.
Free at last!
Thank God almighty, we're free at last!"
We must not allow our creative protest to degenerate into physical violence.
This is the faith that I go back to the mount with.
```

Not only does the summary capture the title of the speech, it captures the main points.

But if you run it again with 10 words per sentence, a lot of the sentences are clearly too long. Because there are only 7 sentences in the whole speech with 10 or fewer words, the program can't honor the input requirements. It defaults to printing the speech from the beginning until the sentence count is at least what was specified in the num_sents variable.

Now, rerun the program and try setting the word count limit to 1,000.

```
Enter max words per sentence for summary: 1000
Enter number of sentences for summary: 15

SUMMARY:
From every mountainside, let freedom ring.
Let freedom ring from Lookout Mountain in Tennessee!
Let freedom ring from every hill and molehill in Mississippi.
Let freedom ring from the curvaceous slopes of California!
Let freedom ring from the snow capped Rockies of Colorado!
But one hundred years later the Negro is still not free.
From the mighty mountains of New York, let freedom ring.
From the prodigious hilltops of New Hampshire, let freedom ring.
And I say to you today my friends, let freedom ring.
I have a dream today.
But not only there; let freedom ring from the Stone Mountain of Georgia!
It is a dream deeply rooted in the American dream.
With this faith we will be able to work together, pray together; to struggle
together, to go to jail together, to stand up for freedom forever, knowing
that we will be free one day.
Free at last!
One hundred years later the life of the Negro is still sadly crippled by the
manacles of segregation and the chains of discrimination.
```

Although longer sentences don't dominate the summary, a few slipped through, making this summary less poetic than the previous one. The lower word count limit forces the previous version to rely more on shorter phrases that act like a chorus.

Project #4: Summarizing Speeches with gensim

In an Emmy award–winning episode of *The Simpsons*, Homer runs for sanitation commissioner using the campaign slogan, "Can't someone else do it?" That's certainly the case with many Python applications: often, when you need to write a script, you learn that someone else has already done it! One example is gensim, an open source library for natural language processing using statistical machine learning.

The word *gensim* stands for "generate similar." It uses a graph-based ranking algorithm called TextRank. This algorithm was inspired by PageRank, invented by Larry Page and used to rank web pages in Google searches. With PageRank, the importance of a website is determined by how many other pages link to it. To use this approach with text processing, algorithms measure how similar each sentence is to all the other sentences. The sentence that is the most like the others is considered the most important.

In this project, you'll use gensim to summarize Admiral William H. McRaven's commencement address, "Make Your Bed," given at the University of Texas at Austin in 2014. This inspirational, 20-minute speech has been viewed more than 10 million times on YouTube and inspired a *New York Times* bestselling book in 2017.

THE OBJECTIVE

Write a Python program that uses the gensim module to summarize a speech.

Installing gensim

The gensim module runs on all the major operating systems but is dependent on NumPy and SciPy. If you don't have them installed, go back to Chapter 1 and follow the instructions in "Installing the Python Libraries" on page 6.

To install gensim on Windows, use pip install -U gensim. To install it in a terminal, use pip install --upgrade gensim. For conda environments, use conda install -c conda-forge gensim. For more on gensim, go to *https://radimrehurek .com/gensim/*.

The Make Your Bed Code

With the *dream_summary.py* program in Project 3, you learned the fundamentals of text extraction. Since you've seen some of the details, use gensim as a streamlined alternative to *dream_summary.py*. Name this new program *bed_summary.py* or download it from the book's website.

Importing Modules, Scraping the Web, and Preparing the Speech String

Listing 3-6 repeats the code used in *dream_summary.py* to prepare the speech as a string. To revisit the detailed code explanation, see page 54.

bed_summary.py, part 1

```
import requests
import bs4
from nltk.tokenize import sent_tokenize
❶ from gensim.summarization import summarize

❷ url = 'https://jamesclear.com/great-speeches/make-your-bed-by-admiral
          -william-h-mcraven'
page = requests.get(url)
page.raise_for_status()
soup = bs4.BeautifulSoup(page.text, 'html.parser')
p_elems = [element.text for element in soup.find_all('p')]

speech = ''.join(p_elems)
```

Listing 3-6: Importing modules and loading the speech as a string

You'll test gensim on the raw speech scraped from the web, so you won't need modules for cleaning the text. The gensim module will also do any counting internally, so you don't need Counter, but you will need gensim's summarize() function to summarize the text ❶. The only other change is to the url address ❷.

Summarizing the Speech

Listing 3-7 completes the program by summarizing the speech and printing the results.

bed_summary.py, part 2

```
print("\nSummary of Make Your Bed speech:")
summary = summarize(speech, word_count=225)
sentences = sent_tokenize(summary)
sents = set(sentences)
print(' '.join(sents))
```

Listing 3-7: Running gensim, removing duplicate lines, and printing the summary

Start by printing a header for your summary. Then, call the gensim summarize() function to summarize the speech in 225 words. This word count will produce about 15 sentences, assuming the average sentence has 15 words. In addition to a word count, you can pass summarize() a ratio, such as ratio=0.01. This will produce a summary whose length is 1 percent of the full document.

Ideally, you could summarize the speech and print the summary in one step.

```
print(summarize(speech, word_count=225))
```

Unfortunately, gensim sometimes duplicates sentences in summaries, and that occurs here:

```
Summary of Make Your Bed speech:
Basic SEAL training is six months of long torturous runs in the soft sand,
midnight swims in the cold water off San Diego, obstacle courses, unending
calisthenics, days without sleep and always being cold, wet and miserable.
Basic SEAL training is six months of long torturous runs in the soft sand,
midnight swims in the cold water off San Diego, obstacle courses, unending
calisthenics, days without sleep and always being cold, wet and miserable.
--snip--
```

To avoid duplicating text, you first need to break out the sentences in the summary variable using the NLTK sent_tokenize() function. Then make a set from these sentences, which will remove duplicates. Finish by printing the results.

Because sets are unordered, the arrangement of the sentences may change if you run the program multiple times.

```
Summary of Make Your Bed speech:
If you can't do the little things right, you will never do the big things
right.And, if by chance you have a miserable day, you will come home to a
bed that is made — that you made — and a made bed gives you encouragement
that tomorrow will be better.If you want to change the world, start off
by making your bed.During SEAL training the students are broken down into
boat crews. It's just the way life is sometimes.If you want to change the
world get over being a sugar cookie and keep moving forward.Every day during
training you were challenged with multiple physical events — long runs, long
swims, obstacle courses, hours of calisthenics — something designed to test
your mettle. Basic SEAL training is six months of long torturous runs in the
soft sand, midnight swims in the cold water off San Diego, obstacle courses,
unending calisthenics, days without sleep and always being cold, wet and
miserable.
>>>
======= RESTART: C:\Python372\sequel\wordcloud\bed_summary.py =======

Summary of Make Your Bed speech:
It's just the way life is sometimes.If you want to change the world get over
being a sugar cookie and keep moving forward.Every day during training you
were challenged with multiple physical events — long runs, long swims,
obstacle courses, hours of calisthenics — something designed to test your
mettle. If you can't do the little things right, you will never do the big
things right.And, if by chance you have a miserable day, you will come home to
a bed that is made — that you made — and a made bed gives you encouragement
that tomorrow will be better.If you want to change the world, start off by
making your bed.During SEAL training the students are broken down into boat
crews. Basic SEAL training is six months of long torturous runs in the soft
sand, midnight swims in the cold water off San Diego, obstacle courses,
unending calisthenics, days without sleep and always being cold, wet and
miserable.
```

If you take the time to read the full speech, you'll probably conclude that gensim produced a fair summary. Although these two results are different, both extracted the key points of the speech, including the reference to making your bed. Given the size of the document, I find this impressive.

Next up, we'll look at a different way of summarizing text using keywords and word clouds.

Project #5: Summarizing Text with Word Clouds

A *word cloud* is a visual representation of text data used to display keyword metadata, called *tags* on websites. In a word cloud, font size or color shows the importance of each tag or word.

Word clouds are useful for highlighting keywords in a document. For example, generating word clouds for each US president's State of the Union address can provide a quick overview of the issues facing the nation that year. In Bill Clinton's first year, the emphasis was on peacetime concerns like healthcare, jobs, and taxes (Figure 3-1).

Figure 3-1: Word cloud made from 1993 State of the Union address by Bill Clinton

Less than 10 years later, George W. Bush's word cloud reveals a focus on security (Figure 3-2).

Figure 3-2: Word cloud made from 2002 State of the Union address by George W. Bush

Another use for word clouds is to extract keywords from customer feedback. If words like *poor, slow,* and *expensive* dominate, you've got a problem! Writers can also use the clouds to compare chapters in a book or scenes in a screenplay. If the author is using very similar language for action scenes and romantic interludes, some editing is needed. If you're a copywriter, clouds can help you check your keyword density for search engine optimization (SEO).

There are lots of ways to generate word clouds, including free websites like *https://www.wordclouds.com/* and *https://www.jasondavies.com/wordcloud/*. But if you want to fully customize your word cloud or embed the generator within another program, you need to do it yourself. In this project, you'll use a word cloud to make a promotional flyer for a school play based on the Sherlock Holmes story *The Hound of the Baskervilles.*

Instead of using the basic rectangle shown in Figures 3-1 and 3-2, you'll fit the words into an outline of Holmes's head (Figure 3-3).

Figure 3-3: Silhouette of Sherlock Holmes

This will make for a more recognizable and eye-catching display.

THE OBJECTIVE

Use the wordcloud module to generate a shaped word cloud for a novel.

The Word Cloud and PIL Modules

You'll use a module called wordcloud to generate the word cloud. You can install it using pip.

```
pip install wordcloud
```

Or, if you're using Anaconda, use the following command:

```
conda install -c conda-forge wordcloud
```

You can find the web page for wordcloud here: *http://amueller.github.io /word_cloud/*.

You'll also need the Python Imaging Library (PIL) to work with images. Use pip again to install it.

```
pip install pillow
```

Or, for Anaconda, use this:

```
conda install -c anaconda pillow
```

In case you're wondering, pillow is the successor project of PIL, which was discontinued in 2011. To learn more about it, visit *https://pillow .readthedocs.io/en/stable/*.

The Word Cloud Code

To make the shaped word cloud, you'll need an image file and a text file. The image shown in Figure 3-3 came from iStock by Getty Images (*https://www.istockphoto.com/vector/detective-hat-gm698950970-129478957/*). This represents the "small" resolution at around 500×600 pixels.

A similar but copyright-free image (*holmes.png*) is provided with the book's downloadable files. You can find the text file (*hound.txt*), image file (*holmes.png*), and code (*wc_hound.py*) in the *Chapter_3* folder.

Importing Modules, Text Files, Image Files, and Stop Words

Listing 3-8 imports modules, loads the novel, loads the silhouette image of Holmes, and creates a set of stop words you'll want to exclude from the cloud.

wc_hound.py, part 1

```
import numpy as np
from PIL import Image
import matplotlib.pyplot as plt
from wordcloud import WordCloud, STOPWORDS

# Load a text file as a string.
❶ with open('hound.txt') as infile:
    text = infile.read()

# Load an image as a NumPy array.
mask = np.array(Image.open('holmes.png'))

# Get stop words as a set and add extra words.
stopwords = STOPWORDS
❷ stopwords.update(['us', 'one', 'will', 'said', 'now', 'well', 'man', 'may',
                  'little', 'say', 'must', 'way', 'long', 'yet', 'mean',
                  'put', 'seem', 'asked', 'made', 'half', 'much',
                  'certainly', 'might', 'came'])
```

Listing 3-8: Importing modules and loading text, image, and stop words

Begin by importing NumPy and PIL. PIL will open the image, and NumPy will turn it into a mask. You started using NumPy in Chapter 1; in case you skipped it, see the "Installing the Python Libraries" section on page 6. Note that the pillow module continues to use the acronym PIL for backward compatibility.

You'll need matplotlib, which you downloaded in the "Installing the Python Libraries" section of Chapter 1, to display the word cloud. The wordcloud module comes with its own list of stop words, so import STOPWORDS along with the cloud functionality.

Next, load the novel's text file and store it in a variable named text ❶. As described in the discussion of Listing 2-2 in Chapter 2, you may encounter a UnicodeDecodeError when loading the text.

```
UnicodeDecodeError: 'ascii' codec can't decode byte 0x93 in position 365:
ordinal not in range(128)
```

In this case, try modifying the open() function by adding encoding and errors arguments.

```
with open('hound.txt', encoding='utf-8', errors='ignore') as infile:
```

With the text loaded, use PIL's Image.open() method to open the image of Holmes and use NumPy to turn it into an array. If you're using the iStock image of Holmes, change the image's filename as appropriate.

Assign the STOPWORDS set imported from wordcloud to the stopwords variable. Then update the set with a list of additional words that you want to exclude ❷. These will be words like *said* and *now* that dominate the word cloud but add no useful content. Determining what they are is an iterative process. You generate the word cloud, remove words that you don't think contribute, and repeat. You can comment out this line to see the benefit.

NOTE   *To update a container like* STOPWORDS, *you need to know whether it's a list, dictionary, set, and so on. Python's built-in* type() *function returns the class type of any object passed as an argument. In this case,* print(type(STOPWORDS)) *yields* <class 'set'>.

Generating the Word Cloud

Listing 3-9 generates the word cloud and uses the silhouette as a *mask*, or an image used to hide portions of another image. The process used by wordcloud is sophisticated enough to fit the words within the mask, rather than simply truncating them at the edges. In addition, numerous parameters are available for changing the appearance of the words within the mask.

wc_hound.py,
part 2

```
wc = WordCloud(max_words=500,
               relative_scaling=0.5,
               mask=mask,
               background_color='white',
               stopwords=stopwords,
               margin=2,
               random_state=7,
```

```
                    contour_width=2,
                    contour_color='brown',
                    colormap='copper').generate(text)

colors = wc.to_array()
```

Listing 3-9: Generating the word cloud

Name a variable `wc` and call `WordCloud()`. There are a lot of parameters, so I've placed each on its own line for clarity. For a list and description of all the parameters available, visit *https://amueller.github.io/word_cloud/generated /wordcloud.WordCloud.html*.

Start by passing the maximum number of words you want to use. The number you set will display the *n* most common words in the text. The more words you choose to display, the easier it will be to define the edges of the mask and make it recognizable. Unfortunately, setting the maximum number too high will also result in a lot of tiny, illegible words. For this project, start with 500.

Next, to control the font size and relative importance of each word, set the `relative_scaling` parameter to 0.5. For example, a value of 0 gives preference to a word's rank to determine the font size, while a value of 1 means that words that occur twice as often will appear twice as large. Values between 0 and 0.5 tend to strike the best balance between rank and frequency.

Reference the `mask` variable and set its background color to `white`. Assigning no color defaults to black. Then reference the `stopwords` set that you edited in the previous listing.

The `margin` parameter will control the spacing of the displayed words. Using 0 will result in tightly packed words. Using 2 will allow for some whitespace padding.

To place the words around the word cloud, use a random number generator and set `random_state` to 7. There's nothing special about this value; I just felt that it produced an attractive arrangement of words.

The `random_state` parameter fixes the seed number so that the results are repeatable, assuming no other parameters are changed. This means the words will always be arranged in the same way. Only integers are accepted.

Now, set `contour_width` to 2. Any value greater than zero creates an outline around a mask. In this case, the outline is squiggly due to the resolution of the image (Figure 3-4).

Set the color of the outline to `brown` using the `contour_color` parameter. Continue using a brownish palette by setting `colormap` to `copper`. In `matplotlib`, a *colormap* is a dictionary that maps numbers to colors. The copper colormap produces text ranging in color from pale flesh to black. You can see its spectrum, along with many other color options, at *https://matplotlib.org/gallery /color/colormap_reference.html*. If you don't specify a colormap, the program will use the default colors.

Use dot notation to call the `generate()` method to build the word cloud. Pass it the `text` string as an argument. End this listing by naming a `colors` variable and calling the `to_array()` method on the `wc` object. This method converts the word cloud image into a `NumPy` array for use with `matplotlib`.

Figure 3-4: Example of masked word cloud with an outline (left) versus without (right)

Plotting the Word Cloud

Listing 3-10 adds a title to the word cloud and uses matplotlib to display it. It also saves the word cloud image as a file.

wc_hound.py, part 3

```
plt.figure()
plt.title("Chamberlain Hunt Academy Senior Class Presents:\n",
          fontsize=15, color='brown')
plt.text(-10, 0, "The Hound of the Baskervilles",
         fontsize=20, fontweight='bold', color='brown')
plt.suptitle("7:00 pm May 10-12 McComb Auditorium",
             x=0.52, y=0.095, fontsize=15, color='brown')
plt.imshow(colors, interpolation="bilinear")
plt.axis('off')
plt.show()
##plt.savefig('hound_wordcloud.png')
```

Listing 3-10: Plotting and saving the word cloud

Start by initializing a matplotlib figure. Then call the title() method and pass it the name of the school, along with a font size and color.

You'll want the name of the play to be bigger and bolder than the other titles. Since you can't change the text style within a string with matplotlib, use the text() method to define a new title. Pass it (*x*, *y*) coordinates (based on the figure axes), a text string, and text style details. Use trial and error with the coordinates to optimize the placement of the text. If you're using the iStock image of Holmes, you may need to change the *x* coordinate from -10 to something else to achieve the best balance with the asymmetrical silhouette.

Finish the titles by placing the play's time and venue at the bottom of the figure. You could use the text() method again, but instead, let's take a look at an alternative, pyplot's suptitle() method. The name stands for "super titles." Pass it the text, the (x, y) figure coordinates, and styling details.

To display the word cloud, call imshow()—for image show—and pass it the colors array you made previously. Specify bilinear for color interpolation.

Turn off the figure axes and display the word cloud by calling show(). If you want to save the figure, uncomment the savefig() method. Note that matplotlib can read the extension in the filename and save the figure in the correct format. As written, the save command will not execute until you manually close the figure.

Fine-Tuning the Word Cloud

Listing 3-10 will produce the word cloud in Figure 3-5. You may get a different arrangement of words as the algorithm is stochastic.

Figure 3-5: The flyer generated by the wc_hound.py code

You can change the size of the display by adding an argument when you initialize the figure. Here's an example: plt.figure(figsize=(50, 60)).

There are many other ways to change the results. For example, setting the margin parameter to 10 yields a sparser word cloud (Figure 3-6).

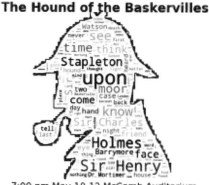

Figure 3-6: The word cloud generated with `margin=10`

Changing the `random_state` parameter will also rearrange the words within the mask (Figure 3-7).

Figure 3-7: The word cloud generated with `margin=10` and `random_state=6`

Tweaking the `max_words` and `relative_scaling` parameters will also change the appearance of the word cloud. Depending on how detail-oriented you are, all this can be a blessing or a curse!

Summary

In this chapter, you used extraction-based summarization techniques to produce a synopsis of Martin Luther King Jr.'s "I Have a Dream" speech. You then used a free, off-the-shelf module called `gensim` to summarize Admiral McRaven's "Make Your Bed" speech with even less code. Finally, you used the `wordcloud` module to create an interesting design with words.

Further Reading

Automate the Boring Stuff with Python: Practical Programming for Total Beginners (No Starch Press, 2015), by Al Sweigart, covers regular expressions in Chapter 7 and web scraping in Chapter 11, including use of the requests and Beautiful Soup modules.

Make Your Bed: Little Things That Can Change Your Life…And Maybe the World, 2nd ed. (Grand Central Publishing, 2017), by William H. McRaven, is a self-help book based on the admiral's commencement address at the University of Texas. You can find the actual speech online on *https://www .youtube.com/*.

Challenge Project: Game Night

Use `wordcloud` to invent a new game for game night. Summarize Wikipedia or IMDb synopses of movies and see whether your friends can guess the movie title. Figure 3-8 shows some examples.

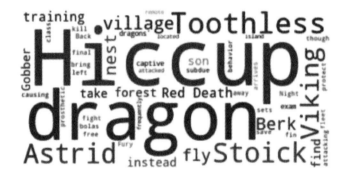

Figure 3-8: Word clouds for two movies released in 2010: How to Train Your Dragon *and* Prince of Persia

If you're not into movies, pick something else. Alternatives include famous novels, *Star Trek* episodes, and song lyrics (Figure 3-9).

Figure 3-9: Word cloud made from song lyrics (Donald Fagen's "I.G.Y.")

Board games have seen a resurgence in recent years, so you could follow this trend and print the word clouds on card stock. Alternatively, you could keep things digital and present the player with multiple-choice answers for each cloud. The game should keep track of the number of correct answers.

Challenge Project: Summarizing Summaries

Test your program from Project 3 on previously summarized text, such as Wikipedia pages. Only five sentences produced a good overview of gensim.

```
Enter max words per sentence for summary: 30
Enter number of sentences for summary: 5

SUMMARY:
Gensim is implemented in Python and Cython.
Gensim is an open-source library for unsupervised topic modeling and natural
language processing, using modern statistical machine learning.
[12] Gensim is commercially supported by the company rare-technologies.com,
who also provide student mentorships and academic thesis projects for Gensim
via their Student Incubator programme.
The software has been covered in several new articles, podcasts and
interviews.
Gensim is designed to handle large text collections using data streaming and
incremental online algorithms, which differentiates it from most other machine
learning software packages that target only in-memory processing.
```

Next, try the gensim version from Project 4 on those boring services agreements no one ever reads. An example Microsoft agreement is available at *https://www.microsoft.com/en-us/servicesagreement/default.aspx*. Of course, to evaluate the results, you'll have to read the full agreement, which almost no one ever does! Enjoy the catch-22!

Challenge Project: Summarizing a Novel

Write a program that summarizes *The Hound of the Baskervilles* by chapter. Keep the chapter summaries short, at around 75 words each.

For a copy of the novel with chapter headings, scrape the text off the Project Gutenberg site using the following line of code: url = 'http://www.gutenberg.org/files/2852/2852-h/2852-h.htm'.

To break out chapter elements, rather than paragraph elements, use this code:

```
chapter_elems = soup.select('div[class="chapter"]')
chapters = chapter_elems[2:]
```

You'll also need to select paragraph elements (p_elems) from within each chapter, using the same methodology as in *dream_summary.py*.

The following snippets show some of the results from using a word count of 75 per chapter:

```
--snip--

Chapter 3:
"Besides, besides—" "Why do you hesitate?" "There is a realm in which the most
acute and most experienced of detectives is helpless." "You mean that the
thing is supernatural?" "I did not positively say so." "No, but you evidently
think it." "Since the tragedy, Mr. Holmes, there have come to my ears several
incidents which are hard to reconcile with the settled order of Nature." "For
example?" "I find that before the terrible event occurred several people had
seen a creature upon the moor which corresponds with this Baskerville demon,
and which could not possibly be any animal known to science.

--snip--

Chapter 6:
"Bear in mind, Sir Henry, one of the phrases in that queer old legend which
Dr. Mortimer has read to us, and avoid the moor in those hours of darkness
when the powers of evil are exalted." I looked back at the platform when we
had left it far behind and saw the tall, austere figure of Holmes standing
motionless and gazing after us.

Chapter 7:
I feared that some disaster might occur, for I was very fond of the old man,
and I knew that his heart was weak." "How did you know that?" "My friend
Mortimer told me." "You think, then, that some dog pursued Sir Charles, and
that he died of fright in consequence?" "Have you any better explanation?" "I
have not come to any conclusion." "Has Mr. Sherlock Holmes?" The words took
away my breath for an instant but a glance at the placid face and steadfast
eyes of my companion showed that no surprise was intended.

--snip--
```

Chapter 14:
"What's the game now?" "A waiting game." "My word, it does not seem a very cheerful place," said the detective with a shiver, glancing round him at the gloomy slopes of the hill and at the huge lake of fog which lay over the Grimpen Mire.

Far away on the path we saw Sir Henry looking back, his face white in the moonlight, his hands raised in horror, glaring helplessly at the frightful thing which was hunting him down.

--snip--

Challenge Project: It's Not Just What You Say, It's How You Say It!

The text summarization programs you have written so far print sentences strictly by their *order of importance*. That means the last sentence in a speech (or any text) might become the first sentence in the summary. The goal of summarization is to find the important sentences, but there's no reason you can't alter the way that they're displayed.

Write a text summarization program that displays the most important sentences in their original *order of appearance*. Compare the results to those produced by the program in Project 3. Does this make a noticeable improvement in the summaries?

4

SENDING SUPER-SECRET MESSAGES WITH A BOOK CIPHER

The Key to Rebecca is a critically acclaimed best-selling novel by Ken Follett. Set in Cairo in World War II and based on actual events, it tells the story of a Nazi spy and the British intelligence officer who pursued him. The title refers to the spy's cipher system, which used the famous gothic novel *Rebecca,* written by Daphne du Maurier, as a key. *Rebecca* is considered one of the greatest novels of the 20th century, and the Germans really did use it as a code book during the war.

The Rebecca cipher is a variation of the *one-time pad,* an unbreakable encryption technique that requires a key that is at least the same size as the message being sent. Both the sender and receiver have a copy of the pad, and after one use, the top sheet is ripped off and discarded.

One-time pads provide absolute, perfect security—uncrackable even by a quantum computer! Despite this, the pads have several practical drawbacks that prevent widespread use. Key among these are the need to securely transport and deliver the pads to the sender and receiver, the need to safely store them, and the difficulty in manually encoding and decoding the messages.

In the *The Key to Rebecca*, both parties must know the encryption rules and have the same edition of the book to use the cipher. In this chapter, you'll transfer the manual method described in the book into a more secure—and easier to use—digital technique. In the process, you'll get to work with useful functions from the Python Standard Library, the collections module, and the random module. You'll also learn a little more about Unicode, a standard used to ensure that characters, such as letters and numbers, are universally compatible across all platforms, devices, and applications.

The One-Time Pad

A one-time pad is basically an ordered stack of sheets printed with truly random numbers, usually in groups of five (Figure 4-1). To make them easy to conceal, the pads tend to be small and may require a powerful magnifying glass to read. Despite being old-school, one-time pads produce the most secure ciphers in the world, as every letter is encrypted with a unique key. As a result, cryptanalysis techniques, such as frequency analysis, simply can't work.

```
73983   91543   74556   01283
24325   88622   92061   02865
22764   47630   14408   80067
13154   81950   11992   84763
46381   99463   49155   40241
98484   77841   03878   14645
11774   73919   83946   40337
12396   26327   76612   12471
18432   41657   93893   10041
77381   39150   47951   83242
 211    02998   15002   08183
```

Figure 4-1: Example of a one-time pad sheet

To encrypt a message with the one-time pad in Figure 4-1, start by assigning each letter of the alphabet a two-digit number. *A* equals 01, *B* equals 02, and so on, as shown in the following table.

A	B	C	D	E	F	G	H	I	J	K	L	M	N	O	P	Q	R	S	T	U	V	W	X	Y	Z
01	02	03	04	05	06	07	08	09	10	11	12	13	14	15	16	17	18	19	20	21	22	23	24	25	26

Next, convert the letters in your short message into numbers:

H	E	R	E		K	I	T	T	Y		K	I	T	T	Y	
08	05	18	05		11	09	20	20	25		11	09	20	20	25	convert letters to numbers

Starting at the upper left of the one-time pad sheet and reading left to right, assign a number pair (*key*) to each letter and add it to the number value of the letter. You'll want to work with base 10 number pairs, so if your sum is greater than 100, use modular arithmetic to truncate the value to the last two digits (103 becomes 03). The numbers in shaded cells in the following diagrams are the result of modular arithmetic.

H	E	R	E		K	I	T	T	Y		K	I	T	T	Y	original message
08	05	18	05		11	09	20	20	25		11	09	20	20	25	convert letters to numbers
73	98	39	15		43	74	55	60	12		83	24	32	58	86	from sender's OTP
81	03	57	20		54	83	75	80	37		94	33	52	78	11	ciphertext

The last row in this diagram represents the ciphertext. Note that KITTY, duplicated in the plaintext, is not repeated in the ciphertext. Each encryption of KITTY is unique.

To decrypt the ciphertext back to plaintext, the recipient uses the same sheet from their identical one-time pad. They place their number pairs below the ciphertext pairs and subtract. When this results in a negative number, they use modular subtraction (adding 100 to the ciphertext value before subtracting). They finish by converting the resulting number pairs back to letters.

81	03	57	20		54	83	75	80	37		94	33	52	78	11	ciphertext
73	98	39	15		43	74	55	60	12		83	24	32	58	86	from recipient's OTP
08	05	18	05		11	09	20	20	25		11	09	20	20	25	convert numbers to letters
H	E	R	E		K	I	T	T	Y		K	I	T	T	Y	decrypted plaintext

To ensure that no keys are repeated, the number of letters in the message can't exceed the number of keys on the pad. This forces the use of short messages, which have the advantage of being easier to encrypt and decrypt and which offer a cryptanalyst fewer opportunities to decipher the message. Some other guidelines include the following:

- Spell out numbers (for example, TWO for 2).

- End sentences with an *X* in place of a period (for example, CALL AT NOONX).

- Spell out any other punctuation that can't be avoided (for example, COMMA).

- End the plaintext message with *XX*.

The Rebecca Cipher

In the novel *The Key to Rebecca*, the Nazi spy uses a variant of the one-time pad. Identical editions of the novel *Rebecca* are purchased in Portugal. Two are retained by the spy; the other two are given to Field Marshal Rommel's staff in North Africa. The encrypted messages are sent by radio on a pre-determined frequency. No more than one message is sent per day and always at midnight.

To use the key, the spy would take the current date—say, May 28, 1942—and add the day to the year (28 + 42 = 70). This would determine which page of the novel to use as a one-time pad sheet. Because May is the fifth month, every fifth word in a sentence would be discounted. Because the Rebecca cipher was meant to be used only during a relatively short period in 1942, the spy didn't have to worry about repetitions in the calendar causing repetition in the keys.

The spy's first message was the following: HAVE ARRIVED. CHECKING IN. ACKNOWLEDGE. Beginning at the top of page 70, he read along until he found the letter *H*. It was the 10th character, discounting every 5th letter. The 10th letter of the alphabet is *J*, so he used this in his ciphertext to represent *H*. The next letter, *A*, was found three letters after *H*, so it was encoded using the third letter of the alphabet, *C*. This continued until the full message was encrypted. For rare letters like *X* or *Z*, author Ken Follett states that special rules were applied but does not describe them.

Using a book in this manner had a distinct advantage over a true one-time pad. To quote Follett, "A pad was unmistakably for the purpose of encipherment, but a book looked quite innocent." A disadvantage remained, however: the process of encryption and decryption is tedious and potentially error prone. Let's see if we can remedy this using Python!

Project #6: The Digital Key to Rebecca

Turning the Rebecca technique into a digital program offers several advantages over a one-time pad:

- The encoding and decoding processes become fast and error-free.
- Longer messages can be sent.
- Periods, commas, and even spaces can be directly encrypted.
- Rare letters, like *z*, can be chosen from anywhere in the book.
- The code book can be hidden among thousands of ebooks on a hard drive or in the cloud.

The last item is important. In the novel, the British intelligence officer finds a copy of *Rebecca* at a captured German outpost. Through simple deductive reasoning he recognizes it as a substitute for a one-time pad. With a digital approach, this would have been much more difficult. In fact, the novel could be kept on a small, easily concealed device such as an SD card.

This would make it similar to a one-time pad, which is often no bigger than a postage stamp.

A digital approach does have one disadvantage, however: the program is a *discoverable* item. Whereas a spy could simply memorize the rules for a one-time pad, with a digital approach the rules must be ensconced in the software. This weakness can be minimized by writing the program so that it looks innocent—or at least cryptic—and having it request input from the user for the message and the name of the code book.

THE OBJECTIVE

Write a Python program that encrypts and decrypts messages using a digital novel as a one-time pad.

The Strategy

Unlike the spy, you won't need all the rules used in the novel, and many wouldn't work anyway. If you've ever used any kind of ebook, you know that page numbers are meaningless. Changes to screen sizes and text sizes render all such page numbers nonunique. And because you can choose letters from anywhere in the book, you don't necessarily need special rules for rare letters or for discounting numbers in a count.

So, you don't need to focus on perfectly reproducing the Rebecca cipher. You just need to produce something similar and, ideally, better.

Luckily, Python *iterables*, such as lists and tuples, use numerical indexes to keep track of every single item within them. By loading a novel as a list, you can use these indexes as unique starting keys for each character. You can then shift the indexes based on the day of the year, emulating the spy's methodology in *The Key to Rebecca*.

Unfortunately, *Rebecca* is not yet in the public domain. In its place, we'll substitute the text file of Sir Arthur Conan Doyle's *The Lost World* that you used in Chapter 2. This novel contains 51 distinct characters that occur 421,545 times, so you can randomly choose indexes with little chance of duplication. This means you can use the whole book as a one-time pad each time you encrypt a message, rather than restrict yourself to a tiny collection of numbers on a single one-time pad sheet.

NOTE *You can download and use a digital version of* Rebecca *if you want. I just can't provide you with a copy for free!*

Because you'll be reusing the book, you'll need to worry about both *message-to-message* and *in-message* duplication of keys. The longer the message, the more material the cryptanalyst can study, and the easier it is to crack the code. And if each message is sent with the same encryption key, all the intercepted messages can be treated as a single large message.

For the message-to-message problem, you can imitate the spy and shift the index numbers by the day of the year, using a range of 1 to 366 to account for leap years. In this scheme, February 1 would be 32. This will effectively turn the book into a new one-time pad sheet each time, as different keys will be used for the same characters. Shifting, by one or more increments, resets all the indexes and essentially "tears off" the previous sheet. And unlike a one-time pad, you don't have to bother with disposing of a piece of paper!

For the in-message duplication problem, you can run a check before transmitting the message. It's unlikely but possible for the program to pick the same letter twice during encryption and thus use the same index twice. Duplicate indexes are basically repeating keys, and these can help a cryptanalyst break your code. So, if duplicate indexes are found, you can rerun the program or reword the message.

You'll also need similar rules to those used in *The Key to Rebecca*.

- Both parties need identical digital copies of *The Lost World*.
- Both parties need to know how to shift the indexes.
- Keep messages as short as possible.
- Spell out numbers.

The Encryption Code

The following *rebecca.py* code will take a message and return an encrypted or plaintext version, as specified by the user. The message can be typed in or downloaded from the book's website. You'll also need the *lost.txt* text file in the same folder as the code.

For clarity, you'll use variable names like *ciphertext, encrypt, message*, and so on. If you were a real spy, however, you'd avoid incriminating terms in case the enemy got their hands on your laptop.

Importing Modules and Defining the main() Function

Listing 4-1 imports modules and defines the main() function, used to run the program. This function will ask for user input, call the functions needed to encrypt or decrypt text, check for duplicate keys, and print the ciphertext or plaintext.

Whether you define main() at the start or end of a program is a matter of choice. Sometimes it provides a good, easily readable summary of the whole program. Other times it may feel out of place, like the cart before the horse. From Python's perspective, it won't matter where you place it so long as you call the function at the end.

rebecca.py,
part 1

```
import sys
import os
import random
from collections import defaultdict, Counter
```

```
def main():
    message = input("Enter plaintext or ciphertext: ")
    process = input("Enter 'encrypt' or 'decrypt': ")
    while process not in ('encrypt', 'decrypt'):
        process = input("Invalid process. Enter 'encrypt' or 'decrypt': ")
    shift = int(input("Shift value (1-366) = "))
    while not 1 <= shift <= 366:
        shift = int(input("Invalid value. Enter digit from 1 to 366: "))
❶   infile = input("Enter filename with extension: ")

    if not os.path.exists(infile):
        print("File {} not found. Terminating.".format(infile), file=sys.stderr)
        sys.exit(1)
    text = load_file(infile)
    char_dict = make_dict(text, shift)

    if process == 'encrypt':
        ciphertext = encrypt(message, char_dict)
❷       if check_for_fail(ciphertext):
            print("\nProblem finding unique keys.", file=sys.stderr)
            print("Try again, change message, or change code book.\n",
                  file=sys.stderr)
            sys.exit()
❸       print("\nCharacter and number of occurrences in char_dict: \n")
        print("{: >10}{: >10}{: >10}".format('Character', 'Unicode', 'Count'))
        for key in sorted(char_dict.keys()):
            print('{:>10}{:>10}{:>10}'.format(repr(key)[1:-1],
                                              str(ord(key)),
                                              len(char_dict[key])))
        print('\nNumber of distinct characters: {}'.format(len(char_dict)))
        print("Total number of characters: {:,}\n".format(len(text)))

        print("encrypted ciphertext = \n {}\n".format(ciphertext))
        print("decrypted plaintext = ")

❹       for i in ciphertext:
            print(text[i - shift], end='', flush=True)

    elif process == 'decrypt':
        plaintext = decrypt(message, text, shift)
        print("\ndecrypted plaintext = \n {}".format(plaintext))
```

Listing 4-1: Importing modules and defining the main() function

Start by importing sys and os, two modules that let you interface with the operating system; then the random module; and then defaultdict and Counter from the collections module.

The collections module is part of the Python Standard Library and includes several container data types. You can use defaultdict to build a dictionary on the fly. If defaultdict encounters a missing key, it will supply a default value rather than throw an error. You'll use it to build a dictionary of the characters in *The Lost World* and their corresponding index values.

A Counter is a dictionary subclass for counting hashable objects. Elements are stored as dictionary keys, and their counts are stored as dictionary values. You'll use this to check your ciphertext and ensure no indexes are duplicated.

At this point, you begin the definition of the main() function. The function starts by asking the user for the message to encrypt or decrypt. For maximum security, the user should type this in. The program then asks the user to specify whether they want encryption or decryption. Once the user chooses, the program asks for the shift value. The shift value represents the day of the year, over the inclusive and consecutive range of 1 to 366. Next, ask for the infile, which will be *lost.txt*, the digital version of *The Lost World* ❶.

Before proceeding, the program checks that the file exists. It uses the operating system module's path.exists() method and passes it the infile variable. If the file doesn't exist or if the path and/or filename is incorrect, the program lets the user know, uses the file=sys.stderr option to color the message "error red" in the Python shell, and terminates the program with sys.exit(1). The 1 is used to flag that the program terminated with an error, as opposed to a clean termination.

Next, you call some functions that you'll define later. The first function loads the *lost.txt* file as a string named text, which includes nonletter characters such as spaces and punctuation. The second builds a dictionary of the characters and their corresponding indexes, with the shift value applied.

Now you start a conditional to evaluate the process being used. As I mentioned, we're using incriminating terms like *encrypt* and *decrypt* for clarity. You'd want to disguise these for real espionage work. If the user has chosen to encrypt, call the function that encrypts the message with the character dictionary. When the function returns, the program has encrypted the message. But don't assume it worked as planned! You need to check that it decrypted correctly and that no keys are duplicated. To do this, you start a series of quality control steps.

First, you check for duplicate keys ❷. If this function returns True, instruct the user to try again, change the message, or change the book to something other than *The Lost World*. For each character in the message, you'll use the char_dict and choose an index at random. Even with hundreds or even thousands of indexes for each character, you may still choose the same index more than once for a given character.

Rerunning the program with slightly different parameters, as listed earlier, should fix this, unless you have a long message with a lot of low-frequency characters. Handling this rare case may require rewording the message or finding a larger manuscript than *The Lost World*.

NOTE *Python's random module does not produce truly random numbers but rather pseudorandom numbers that can be predicted. Any cipher using pseudorandom numbers can potentially be cracked by a cryptanalyst. For maximum security when generating random numbers, you should use Python's os.urandom() function.*

Now, print the contents of the character dictionary so you can see how many times the various characters occur in the novel ❸. This will help guide what you put in messages, though *The Lost World* contains a healthy helping of useful characters.

Character and number of occurrences in char_dict:

Character	Unicode	Count
\n	10	7865
	32	72185
!	33	282
"	34	2205
'	39	761
(	40	62
)	41	62
,	44	5158
-	45	1409
.	46	3910
0	48	1
1	49	7
2	50	3
3	51	2
4	52	2
5	53	2
6	54	1
7	55	4
8	56	5
9	57	2
:	58	41
;	59	103
?	63	357
a	97	26711
b	98	4887
c	99	8898
d	100	14083
e	101	41156
f	102	7705
g	103	6535
h	104	20221
i	105	21929
j	106	431
k	107	2480
l	108	13718
m	109	8438
n	110	21737
o	111	25050
p	112	5827
q	113	204
r	114	19407
s	115	19911
t	116	28729
u	117	10436
v	118	3265

```
w      119      8536
x      120       573
y      121      5951
z      122       296
{      123         1
}      125         1

Number of distinct characters: 51
Total number of characters: 421,545
```

To generate this table, you use Python's Format Specification Mini-Language (*https://docs.python.org/3/library/string.html#formatspec*) to print headers for the three columns. The number in curly brackets denotes how many characters should be in the string, and the greater-than sign designates right justification.

The program then loops through the keys in the character dictionary and prints them using the same column width and justification. It prints the character, its Unicode value, and the number of times it occurs in the text.

You use repr() to print the key. This built-in function returns a string containing a printable representation of an object. That is, it returns all information about the object in a format useful for debugging and development purposes. This allows you to explicitly print characters like newline (\n) and space. The index range [1:-1] excludes the quotes on both sides of the output string.

The ord() built-in function returns an integer representing the Unicode code point for a character. Computers deal only with numbers, so they must assign a number to every possible character, such as %, 5, ☺, or *A*. The *Unicode Standard* ensures that every character, no matter what platform, device, application, or language, has a unique number and is universally compatible. By showing the user the Unicode values, the program lets the user pick up on anything strange happening with a text file, such as the same letter showing up as multiple distinct characters.

For the third column, you get the length of each dictionary key. This will represent the number of times that character appears in the novel. The program then prints the number of distinct characters and the total of all characters in the text.

Finally, you finish the encryption process by printing the ciphertext, and then the decrypted plaintext, as a check. To decipher the message, the program loops through each item in the ciphertext and uses the item as an index for text ❹, subtracting the shift value, which was added earlier. When you print the results, the program uses end='' in place of the default newline, so each character isn't on a separate line.

The main() function ends with a conditional statement to check whether process == 'decrypt'. If the user chooses to decrypt the message, the program calls the decrypt() function and then prints the decrypted plaintext. Note that you could simply use else here, but I chose to use elif for clarity and readability.

Loading a File and Making a Dictionary

Listing 4-2 defines functions to load a text file and make a dictionary of characters in the file and their corresponding indexes.

rebecca.py, part 2

```
def load_file(infile):
    """Read and return text file as a string of lowercase characters."""
    with open(infile) as f:
        loaded_string = f.read().lower()
    return loaded_string

❶ def make_dict(text, shift):
    """Return dictionary of characters and shifted indexes."""
    char_dict = defaultdict(list)
    for index, char in enumerate(text):
      ❷ char_dict[char].append(index + shift)
    return char_dict
```

Listing 4-2: Defining the load_file() *and* make_dict() *functions*

This listing begins by defining a function to load a text file as a string. Using with to open the file ensures it will close automatically when the function ends.

Some users may get an error, such as the following one, when loading text files:

```
UnicodeDecodeError: 'charmap' codec can't decode byte 0x81 in position
27070:character maps to <undefined>
```

In this case, try modifying the open function by adding the encoding and errors arguments.

```
with open(infile, encoding='utf-8', errors='ignore') as f:
```

For more on this issue, see page 35 in Chapter 2.

After opening the file, read it to a string and convert all the text to lower-case. Then return the string.

The next step is to turn the string into a dictionary. Define a function that takes this string and the shift value as arguments ❶. The program creates a char_dict variable using defaultdict(). This variable will be a dictionary. The program then passes the type constructor for list to defaultdict(), as you want the dictionary values to be a list of indexes.

With defaultdict(), whenever an operation encounters an item that isn't already in the dictionary, a function named default_factory() is called with no arguments, and the output is used as the value. Any key that doesn't exist gets the value returned by default_factory, and no KeyError is raised.

If you tried to make the dictionary on the fly without the handy collections module, you'd get the KeyError, as shown in the next example.

```
>>> mylist = ['a', 'b', 'c']
>>> d = dict()
>>> for index, char in enumerate(mylist):
    d[char].append(index)

Traceback (most recent call last):
 File "<pyshell#16>", line 2, in <module>
  d[char].append(index)
KeyError: 'a'
```

The built-in enumerate() function acts as an automatic counter, so you can easily get the index for each character in the string derived from *The Lost World*. The keys in char_dict are characters, and the characters can occur thousands of times within text. So, the dictionary values are lists that hold the indexes for all these character occurrences. By adding the shift value to the index when it is appended to a value list, you ensure that the indexes will be unique for each message ❷.

Finish the function by returning the character dictionary.

Encrypting the Message

Listing 4-3 defines a function to encrypt the message. The resulting ciphertext will be a list of indexes.

rebecca.py, part 3

```
def encrypt(message, char_dict):
    """Return list of indexes representing characters in a message."""
    encrypted = []
    for char in message.lower():
      ❶ if len(char_dict[char]) > 1:
            index = random.choice(char_dict[char])
        elif len(char_dict[char]) == 1: # Random.choice fails if only 1 choice
            index = char_dict[char][0]
      ❷ elif len(char_dict[char]) == 0:
            print("\nCharacter {} not in dictionary.".format(char),
                    file=sys.stderr)
            continue
        encrypted.append(index)
    return encrypted
```

Listing 4-3: Defining a function to encrypt the plaintext message

The encrypt() function will take the message and char_dict as arguments. Start it by creating an empty list to hold the ciphertext. Next, start looping through the characters in message and converting them to lowercase to match the characters in char_dict.

If the number of indexes associated with the character is greater than 1, the program uses the random.choice() method to choose one of the character's indexes at random ❶.

If a character occurs only once in char_dict, random.choice() will throw an error. To handle this, the program uses a conditional and hardwires the choice of the index, which will be at position [0].

If the character doesn't exist in *The Lost World*, it won't be in the dictionary, so use a conditional to check for this ❷. If it evaluates to True, print an alert for the user and use continue to return to the start of the loop without choosing an index. Later, when the quality control steps run on the ciphertext, a space will appear in the decrypted plaintext where this character should be.

If continue is not called, then the program appends the index to the encrypted list. When the loop ends, you return the list to end the function.

To see how this works, let's look at the first message the Nazi spy sends in *The Key to Rebecca*, shown here:

HAVE ARRIVED. CHECKING IN. ACKNOWLEDGE.

Using this message and a shift value of 70 yielded the following randomly generated ciphertext:

```
[125711, 106950, 85184, 43194, 45021, 129218, 146951, 157084, 75611, 122047,
121257, 83946, 27657, 142387, 80255, 160165, 8634, 26620, 105915, 135897,
22902, 149113, 110365, 58787, 133792, 150938, 123319, 38236, 23859, 131058,
36637, 108445, 39877, 132085, 86608, 65750, 10733, 16934, 78282]
```

Your results may differ due to the stochastic nature of the algorithm.

Decrypting the Message

Listing 4-4 defines a function to decrypt the ciphertext. The user will copy and paste the ciphertext when asked for input by the main() function.

rebecca.py, part 4

```
def decrypt(message, text, shift):
    """Decrypt ciphertext list and return plaintext string."""
    plaintext = ''
    indexes = [s.replace(',', '').replace('[', '').replace(']', '')
               for s in message.split()]
    for i in indexes:
        plaintext += text[int(i) - shift]
    return plaintext
```

Listing 4-4: Defining a function to decrypt the plaintext message

The listing starts by defining a function named decrypt() with the message, the novel (text), and the shift value as parameters. Of course, the message will be in ciphertext form, consisting of a list of numbers representing shifted indexes. You immediately create an empty string to hold the decrypted plaintext.

Most people will copy and paste the ciphertext when prompted for input by the main() function. This input may or may not contain the square brackets that came with the list. And because the user entered the ciphertext using the input() function, the results are a *string*. To convert the indexes to integers that can be shifted, you first need to remove the non-digit characters. Do this using the string replace() and split() methods, while also using list comprehension to return a list. List comprehension is a shorthand way to execute loops in Python.

To use replace(), you pass it the character you want replaced followed by the character used to replace it. In this case, use a space for the replacement. Note that you can "string" these together with dot notation, handling the commas and brackets all in one go. How cool is that?

Next, start looping through the indexes. The program converts the current index from a string to an integer so you can subtract the shift value that was applied during encryption. You use the index to access the character list and get the corresponding character. Then you add the character to the plaintext string and return plaintext when the loop ends.

Checking for Failure and Calling the main() Function

Listing 4-5 defines a function to check the ciphertext for duplicate indexes (keys) and finishes the program by calling the main() function. If the function discovers duplicate indexes, the encryption might have been compromised, and the main() function will tell the user how to fix it before terminating.

rebecca.py,
part 5

```
def check_for_fail(ciphertext):
    """Return True if ciphertext contains any duplicate keys."""
    check = [k for k, v in Counter(ciphertext).items() if v > 1]
    if len(check) > 0:
        return True

if __name__ == '__main__':
    main()
```

Listing 4-5: Defining a function to check for duplicate indexes and calling main()

This listing defines a function named check_for_fail() that takes the ciphertext as an argument. It checks to see whether any of the indexes in the ciphertext are repeated. Remember, the one-time pad approach works because every key is unique; thus, every index in the ciphertext should be unique.

To look for repeats, the program uses Counter again. It employs list comprehension to build a list containing all the duplicate indexes. Here, k stands for (dictionary) key, and v stands for (dictionary) value. Since Counter produces a dictionary of counts for each key, what you're saying here is this: For every key-value pair in a dictionary made from the ciphertext, create a list of all the keys that occur more than once. If there are duplicates, append the corresponding key to the check list.

Now all you need to do is get the length of check. If it is greater than zero, the encryption is compromised, and the program returns True.

The program ends with the boilerplate code to call the program as a module or in stand-alone mode.

Sending Messages

The following message is based on a passage from *The Key to Rebecca.* You can find it in the downloadable *Chapter_4* folder as *allied_attack_plan.txt.*

As a test, try sending it with a shift of 70. Use your operating system's Select All, Copy, and Paste commands to transfer the text when asked for input. If it doesn't pass the check_for_fail() test, run it again!

```
Allies plan major attack for Five June. Begins at oh five twenty with
bombardment from Aslagh Ridge toward Rommel east flank. Followed by tenth
Indian Brigade infantry with tanks of twenty second Armored Brigade on Sidi
Muftah. At same time, thirty second Army Tank Brigade and infantry to charge
north flank at Sidra Ridge. Three hundred thirty tanks deployed to south and
seventy to north.
```

The nice thing about this technique is that you can use proper punctuation, at least if you type the message into the interpreter window. Text copied in from outside may need to be stripped of the newline character (such as \r\n or \n), placed wherever the carriage return was used.

Of course, only characters that occur in *The Lost World* can be encrypted. The program will warn you of exceptions and then replace missing characters with a space.

To be sneaky, you don't want to save plaintext or ciphertext messages to a file. Cutting and pasting from the shell is the way to go. Just remember to copy something new when you're finished so you don't leave incriminating evidence on your clipboard!

If you want to get fancy, you can copy and paste text to the clipboard straight from Python using pyperclip, written by Al Sweigart. You can learn more at *https://pypi.org/project/pyperclip/*.

Summary

In this chapter, you got to work with defaultdict and Counter from the collections module; choice() from the random module; and replace(), enumerate(), ord(), and repr() from the Python Standard Library. The result was an encryption program, based on the one-time pad technique, that produces unbreakable ciphertext.

Further Reading

The Key to Rebecca (Penguin Random House, 1980), by Ken Follett, is an exciting novel noted for its depth of historical detail, accurate descriptions of Cairo in World War II, and thrilling espionage storyline.

The Code Book: The Science of Secrecy from Ancient Egypt to Quantum Cryptography (Anchor, 2000), by Simon Singh, is an interesting review of cryptography through the ages, including a discussion of the one-time pad.

If you enjoy working with ciphers, check out *Cracking Codes with Python* (No Starch Press, 2018), by Al Sweigart. Aimed at beginners in both cryptography and Python programming, this book covers many cipher types, including reverse, Caesar, transposition, substitution, affine, and Vigenère.

Impractical Python Projects: Playful Programming Activities to Make You Smarter (No Starch Press, 2019), by Lee Vaughan, includes additional ciphers such as the Union route cipher, the rail fence cipher, and the Trevanion null cipher as well as a technique for writing with invisible electronic ink.

Practice Project: Charting the Characters

If you have `matplotlib` installed (see "Installing the Python Libraries" on page 6), you can visually represent the available characters in *The Lost World*, along with their frequency of occurrence, using a bar chart. This can complement the shell printout of each character and its count currently used in the *rebecca.py* program.

The internet is rife with example code for `matplotlib` plots, so just search for *make a simple bar chart matplotlib*. You'll want to sort the counts in descending order before plotting.

The mnemonic for remembering the most common letters in English is "etaoin." If you plot in descending order, you'll see that *The Lost World* dataset is no exception (Figure 4-2)!

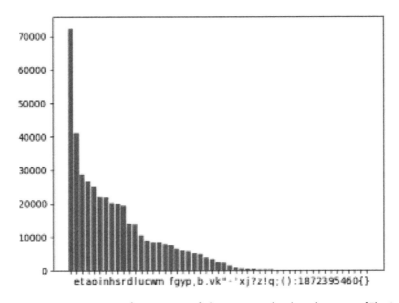

Figure 4-2: Frequency of occurrence of characters in the digital version of The Lost World

Note that the most common character is a space. This makes it easy to encrypt spaces, further confounding any cryptanalysis!

You can find a solution, *practice_barchart.py*, in the appendix and on the book's website.

Practice Project: Sending Secrets the WWII Way

According to the Wikipedia article on *Rebecca* (*https://en.wikipedia.org/wiki/Rebecca_(novel)*), the Germans in North Africa in World War II really did attempt to use the novel as the key to a book code. Rather than encode the message letter by letter, sentences would be made using single words in the book, referred to by page number, line, and position in the line.

Copy and edit the *rebecca.py* program so that it uses words rather than letters. To get you started, here's how to load the text file as a list of words, rather than characters, using list comprehension:

```
with open('lost.txt') as f:
    words = [word.lower() for line in f for word in line.split()]
    words_no_punct = ["".join(char for char in word if char.isalpha())
                for word in words]

print(words_no_punct[:20])  # Print first 20 words as a QC check
```

The output should look like this:

```
['i', 'have', 'wrought', 'my', 'simple', 'plan', 'if', 'i', 'give', 'one',
'hour', 'of', 'joy', 'to', 'the', 'boy', 'whos', 'half', 'a', 'man']
```

Note that all punctuation, including apostrophes, has been removed. Messages will need to follow this convention.

You'll also need to handle words, such as proper names and place names, that don't occur in *The Lost World*. One approach would be a "first-letter mode" where the recipient uses only the first letter of each word between flags. The flags should be commonly occurring words, like *a* and *the*, doubled. Alternate their use to make it easier to identify the start and end flags. In this case, *a a* indicates the start of first-letter mode, and *the the* indicates the end. For example, to handle the phrase *Sidi Muftah with ten tanks*, first run it straight up to identify missing words.

```
Enter plaintext or ciphertext: sidi muftah with ten tanks
Enter 'encrypt' or 'decrypt': encrypt
Shift value (1-365) = 5
Enter filename with extension: lost.txt

Character sidi not in dictionary.

Character muftah not in dictionary.

Character tanks not in dictionary.

encrypted ciphertext =
 [23371, 7491]

decrypted plaintext =
with ten
```

With the missing words identified, reword the message to spell them using first-letter mode. I've highlighted the first letters in gray in the following snippet:

```
Enter plaintext or ciphertext: a a so if do in my under for to all he the the
with ten a a tell all night kind so the the
Enter 'encrypt' or 'decrypt': encrypt
Shift value (1-365) = 5
Enter filename with extension: lost.txt

encrypted ciphertext =
 [29910, 70641, 30556, 60850, 72292, 32501, 6507, 18593, 41777, 23831, 41833,
16667, 32749, 3350, 46088, 37995, 12535, 30609, 3766, 62585, 46971, 8984,
44083, 43414, 56950]

decrypted plaintext =
a a so if do in my under for to all he the the with ten a a tell all night
kind so the the
```

There are 1,864 occurrences of *a* and 4,442 of *the* in *The Lost World*. If you stick to short messages, you shouldn't duplicate keys. Otherwise, you may need to use multiple flag characters or disable the check-for-fail() function and accept some duplicates.

Feel free to come up with your own method for handling problem words. As consummate planners, the Germans surely had *something* in mind or they wouldn't have considered a book code in the first place!

You can find a simple first-letter solution, *practice_WWII_words.py*, in the appendix or online at *https://nostarch.com/real-world-python/*.

5

FINDING PLUTO

According to Woody Allen, 80 percent of success is just showing up. This certainly describes the success of Clyde Tombaugh, an untrained Kansas farm boy growing up in the 1920s. With a passion for astronomy but no money for college, he took a stab in the dark and mailed his best astronomical sketches to Lowell Observatory. To his great surprise, he was hired as an assistant. A year later, he had discovered Pluto and gained eternal glory!

Percival Lowell, the famous astronomer and founder of Lowell Observatory, had postulated the presence of Pluto based on perturbations in the orbit of Neptune. His calculations were wrong, but by pure coincidence, he correctly predicted Pluto's orbital path. Between 1906 and his death in 1916, he had photographed Pluto twice. Both times, his team failed to notice it. Tombaugh, on the other hand, photographed *and* recognized Pluto in January 1930, after only a year of searching (Figure 5-1).

January 23, 1930   January 29, 1930

Figure 5-1: Discovery plates for Pluto, indicated by the arrow

What Tombaugh accomplished was extraordinary. Without computers, the methodology he followed was impractical, tedious, and demanding. He had to photograph and re-photograph small parts of the sky night after night, usually in a freezing cold dome shaken by icy winds. He then developed and sifted through all the negatives, searching for the faintest signs of movement within crowded star fields.

Although he lacked a computer, he did have a state-of-the-art device, known as a *blink comparator*, that let him rapidly switch between negatives from successive nights. As viewed through the blink comparator, the stars remained stationary, but Pluto, a moving object, flashed on and off like a beacon.

In this chapter, you'll first write a Python program that replicates an early 20th-century blink comparator. Then you'll move into the 21st century and write a program that automates the detection of moving objects using modern computer vision techniques.

NOTE   *In 2006, the International Astronomical Union reclassified Pluto as a dwarf planet. This was based on the discovery of other Pluto-sized bodies in the Kuiper Belt, including one—Eris—that is volumetrically smaller but 27 percent more massive than Pluto.*

Project #7: Replicating a Blink Comparator

Pluto may have been photographed with a telescope, but it was found with a microscope. The blink comparator (Figure 5-2), also called the *blink microscope*, lets the user mount two photographic plates and rapidly switch from looking at one to the other. During this "blinking," any object that changes position between photographs will appear to jump back and forth.

Figure 5-2: A blink comparator

For this technique to work, the photos need to be taken with the same exposure and under similar viewing conditions. Most importantly, the stars in the two images must line up perfectly. In Tombaugh's day, technicians achieved this through painstaking manual labor; they carefully guided the telescope during the hour-long exposures, developed the photographic plates, and then shifted them in the blink comparator to fine-tune the alignment. Because of this exacting work, it would sometimes take Tombaugh a week to examine a single pair of plates.

In this project, you'll digitally duplicate the process of aligning the plates and blinking them on and off. You'll work with bright and dim objects, see the impact of different exposures between photos, and compare the use of positive images to the negative ones that Tombaugh used.

THE OBJECTIVE

Write a Python program that aligns two nearly identical images and displays each one in rapid succession in the same window.

The Strategy

The photos for this project are already taken, so all you need to do is align them and flash them on and off. Aligning images is often referred to as image *registration*. This involves making a combination of vertical, horizontal, or rotational transformations to one of the images. If you've ever taken a panorama with a digital camera, you've seen registration at work.

Image registration follows these steps:

1. Locate distinctive features in each image.
2. Numerically describe each feature.

3. Use the numerical descriptors to match identical features in each image.

4. Warp one image so that matched features share the same pixel locations in both images.

For this to work well, the images should be the same size and cover close to the same area.

Fortunately, the OpenCV Python package ships with algorithms that perform these steps. If you skipped Chapter 1, you can read about OpenCV on page 6.

Once the images are registered, you'll need to display them in the same window so that they overlay exactly and then loop through the display a set number of times. Again, you can easily accomplish this with the help of OpenCV.

The Data

The images you'll need are in the *Chapter_5* folder in the book's supporting files, downloadable from *https://nostarch.com/real-world-python/*. The folder structure should look like Figure 5-3. After downloading the folders, don't change this organizational structure or the folder contents and names.

Figure 5-3: The folder structure for Project 7

The *night_1* and *night_2* folders contain the input images you'll use to get started. In theory, these would be images of the same region of space taken on different nights. The ones used here are the same star field image to which I've added an artificial *transient*. A transient, short for *transient astronomical event*, is a celestial object whose motion is detectable over relatively short time frames. Comets, asteroids, and planets can all be considered transients, as their movement is easily detected against the more static background of the galaxy.

Table 5-1 briefly describes the contents of the *night_1* folder. This folder contains files with *left* in their filenames, which means they should go on the left side of a blink comparator. The images in the *night_2* folder contain *right* in the filenames and should go on the other side.

Table 5-1: Files in the *night_1* folder

Filename	Description
1_bright_transient_left.png	Contains a large, bright transient
2_dim_transient_left.png	Contains a dim transient a single pixel in diameter
3_diff_exposures_left.png	Contains a dim transient with an overexposed background
4_single_transient_left.png	Contains a bright transient in left image only
5_no_transient_left.png	Star field with no transient
6_bright_transient_neg_left.png	A negative of the first file to show the type of image Tombaugh used

Figure 5-4 is an example of one of the images. The arrow points to the transient (but isn't part of the image file).

Figure 5-4: 1_bright_transient_left.png *with an arrow indicating the transient*

To duplicate the difficulty in perfectly aligning a telescope from night to night, I've slightly shifted the images in the *night_2* folder with respect to those in *night_1*. You'll need to loop through the contents of the two folders, registering and comparing each pair of photos. For this reason, the number of files in each folder should be the same, and the naming convention should ensure that the photos are properly paired.

The Blink Comparator Code

The following *blink_comparator.py* code will digitally duplicate a blink comparator. Find this program in the *Chapter_5* folder from the website. You'll also need the folders described in the previous section. Keep the code in the folder above the *night_1* and *night_2* folders.

Importing Modules and Assigning a Constant

Listing 5-1 imports the modules you'll need to run the program and assigns a constant for the minimum number of keypoint matches to accept. Also called interest points, *keypoints* are interesting features in an image that you can use to characterize the image. They're usually associated with sharp changes in intensity, such as corners or, in this case, stars.

*blink
_comparator.py,
part 1*

```
import os
from pathlib import Path
import numpy as np
import cv2 as cv

MIN_NUM_KEYPOINT_MATCHES = 50
```

Listing 5-1: Importing modules and assigning a constant for keypoint matches

Start by importing the operating system module, which you'll use to list the contents of folders. Then import `pathlib`, a handy module that simplifies working with files and folders. Finish by importing `NumPy` and `cv` (OpenCV) for working with images. If you skipped Chapter 1, you can find installation instructions for `NumPy` on page 8.

Assign a constant variable for the minimum number of keypoint matches to accept. For efficiency, you ideally want the smallest value that will yield an acceptable registration result. In this project, the algorithm runs so quickly that you can increase this value without a significant cost.

Defining the main() Function

Listing 5-2 defines the first part of the `main()` function, used to run the program. These initial steps create lists and directory paths used to access the various image files.

*blink
_comparator.py,
part 2*

```
def main():
    """Loop through 2 folders with paired images, register & blink images."""
    night1_files = sorted(os.listdir('night_1'))
    night2_files = sorted(os.listdir('night_2'))
    path1 = Path.cwd() / 'night_1'
    path2 = Path.cwd() / 'night_2'
    path3 = Path.cwd() / 'night_1_registered'
```

Listing 5-2: Defining the first part of main(), used to manipulate files and folders

Start by defining `main()` and then use the os module's `listdir()` method to create a list of the filenames in the *night_1* and *night_2* folders. For the *night_1* folder, `listdir()` returns the following:

```
['1_bright_transient_left.png', '2_dim_transient_left.png', '3_diff_exposures_
left.png', '4_no_transient_left.png', '5_bright_transient_neg_left.png']
```

Note that `os.listdir()` does not impose an order on the files when they're returned. The underlying operating system determines the order, meaning macOS will return a different list than Windows! To ensure that the lists are consistent and the files are paired correctly, wrap os.listdir() with the built-in sorted() function. This function will return the files in numerical order, based on the first character in the filename.

Next, assign path names to variables using the pathlib Path class. The first two variables will point to the two input folders, and the third will point to an output folder to hold the registered images.

The pathlib module, introduced in Python 3.4, is an alternative to os.path for handling file paths. The os module treats paths as strings, which can be cumbersome and requires you to use functionality from across the Standard Library. Instead, the pathlib module treats paths as objects and gathers the necessary functionality in one place. The official documentation for pathlib is at *https://docs.python.org/3/library/pathlib.html*.

For the first part of the directory path, use the cwd() class method to get the current working directory. If you have at least one Path object, you can use a mix of objects and strings in the path designation. You can join the string, representing the folder name, with the / symbol. This is similar to using os.path.join(), if you're familiar with the os module.

Note that you will need to execute the program from within the project directory. If you call it from elsewhere in the filesystem, it will fail.

Looping in main()

Listing 5-3, still in the main() function, runs the program with a big for loop. This loop will take a file from each of the two "night" folders, load them as grayscale images, find matching keypoints in each image, use the keypoints to warp (or *register*) the first image to match the second, save the registered image, and then compare (or *blink*) the registered first image with the original second image. I've also included a few optional quality control steps that you can comment out once you're satisfied with the results.

*blink
_comparator.py,
part 3*

```
for i, _ in enumerate(night1_files):
    img1 = cv.imread(str(path1 / night1_files[i]), cv.IMREAD_GRAYSCALE)
    img2 = cv.imread(str(path2 / night2_files[i]), cv.IMREAD_GRAYSCALE)
    print("Comparing {} to {}.\n".format(night1_files[i], night2_files[i]))
❶  kp1, kp2, best_matches = find_best_matches(img1, img2)
    img_match = cv.drawMatches(img1, kp1, img2, kp2,
                               best_matches, outImg=None)
    height, width = img1.shape
    cv.line(img_match, (width, 0), (width, height), (255, 255, 255), 1)
❷  QC_best_matches(img_match)  # Comment out to ignore.
    img1_registered = register_image(img1, img2, kp1, kp2, best_matches)

❸  blink(img1, img1_registered, 'Check Registration', num_loops=5)
    out_filename = '{}_registered.png'.format(night1_files[i][:-4])
    cv.imwrite(str(path3 / out_filename), img1_registered) # Will overwrite!
    cv.destroyAllWindows()
    blink(img1_registered, img2, 'Blink Comparator', num_loops=15)
```

Listing 5-3: Running the program loop in main()

Begin the loop by enumerating the `night1_files` list. The `enumerate()` built-in function adds a counter to each item in the list and returns this counter along with the item. Since you only need the counter, use a single underscore (_) for the list item. By convention, the single underscore indicates a temporary or insignificant variable. It also keeps code-checking programs, such as Pylint, happy. Were you to use a variable name here, such as `infile`, Pylint would complain about an *unused variable*.

```
W: 17,11: Unused variable 'infile' (unused-variable)
```

Next, load the image, along with its pair from the `night2_files` list, using OpenCV. Note that you have to convert the path to a string for the `imread()` method. You'll also want to convert the image to grayscale. This way, you'll need to work with only a single channel, which represents intensity. To keep track of what's going on during the loop, print a message to the shell indicating which files are being compared.

Now, find the keypoints and their best matches ❶. The `find_best_matches()` function, which you'll define later, will return these values as three variables: `kp1`, which represents the keypoints for the first loaded image; `kp2`, which represents the keypoints for the second; and `best_matches`, which represents a list of the matching keypoints.

So you can visually check the matches, draw them on `img1` and `img2` using OpenCV's `drawMatches()` method. As arguments, this method takes each image with its keypoints, the list of best matching keypoints, and an output image. In this case, the output image argument is set to `None`, as you're just going to look at the output, not save it to a file.

To distinguish between the two images, draw a vertical white line down the right side of `img1`. First get the height and width of the image using `shape`. Next, call OpenCV's `line()` method and pass it the image on which you want to draw, the start and end coordinates, the line color, and the thickness. Note that this is a color image, so to represent white, you need the full BGR tuple (255, 255, 255) rather than the single intensity value (255) used in grayscale images.

Now, call the quality control function—which you'll define later—to display the matches ❷. Figure 5-5 shows an example output. You may want to comment out this line after you confirm the program is behaving correctly.

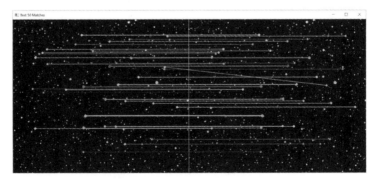

Figure 5-5: Example output of the `QC_best_matches()` function

With the best keypoint matches found and checked, it's time to register the first image to the second. Do this with a function you'll write later. Pass the function the two images, the keypoints, and the list of best matches.

The blink comparator, named blink(), is another function that you'll write later. Call it here to see the effect of the registration process on the first image. Pass it the original and registered images, a name for the display window, and the number of blinks you want to perform ❸. The function will flash between the two images. The amount of "wiggle" you see will depend on the amount of warping needed to match img2. This is another line you may want to comment out after you've confirmed that the program runs as intended.

Next, save the registered image into a folder named *night_1_registered*, which the path3 variable points to. Start by assigning a filename variable that references the original filename, with *_registered.png* appended to the end. So you don't repeat the file extension in the name, use index slicing ([:-4]) to remove it before adding the new ending. Finish by using imwrite() to save the file. Note that this will overwrite existing files with the same name without warning.

You'll want an uncluttered view when you start looking for transients, so call the method to destroy all the current OpenCV windows. Then call the blink() function again, passing it the registered image, the second image, a window name, and the number of times to loop through the images. The first images are shown side by side in Figure 5-6. Can you find the transient?

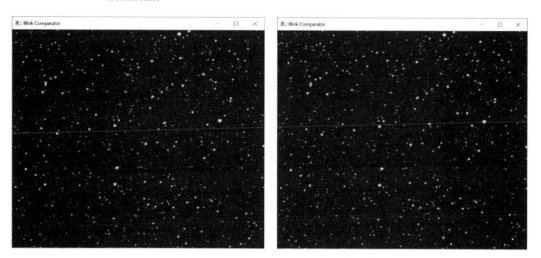

Figure 5-6: Blink Comparator windows for first image in night_1_registered *and* night_2 *folders*

Finding the Best Keypoint Matches

Now it's time to define the functions used in main(). Listing 5-4 defines the function that finds the best keypoint matches between each pair of images extracted from the *night_1* and *night_2* folders. It should locate, describe, and match keypoints, generate a list of the matches, and then truncate that list by the constant for the minimum number of acceptable keypoints. The function returns the list of keypoints for each image and the list of best matches.

*blink
_comparator.py,*
part 4

```
def find_best_matches(img1, img2):
    """Return list of keypoints and list of best matches for two images."""
    orb = cv.ORB_create(nfeatures=100)  # Initiate ORB object.
❶ kp1, desc1 = orb.detectAndCompute(img1, mask=None)
    kp2, desc2 = orb.detectAndCompute(img2, mask=None)
    bf = cv.BFMatcher(cv.NORM_HAMMING, crossCheck=True)
❷ matches = bf.match(desc1, desc2)
    matches = sorted(matches, key=lambda x: x.distance)
    best_matches = matches[:MIN_NUM_KEYPOINT_MATCHES]

    return kp1, kp2, best_matches
```

Listing 5-4: Defining the function to find the best keypoint matches

Start by defining the function, which takes two images as arguments. The main() function will pick these images from the input folders with each run of the for loop.

Next, create an orb object using OpenCV's ORB_create() method. ORB is an acronym of nested acronyms: *O*riented *F*AST and *R*otated *B*RIEF.

FAST, short for *F*eatures from *A*ccelerated *S*egment *T*est, is a fast, efficient, and free algorithm for *detecting* keypoints. To *describe* the keypoints so that you can compare them across different images, you need BRIEF. Short for *B*inary *R*obust *I*ndependent *E*lementary *F*eatures, BRIEF is also fast, compact, and open source.

ORB combines FAST and BRIEF into a matching algorithm that works by first detecting distinctive regions in an image, where pixel values change sharply, and then recording the position of these distinctive regions as *keypoints*. Next, ORB describes the feature found at the keypoint using numerical arrays, or *descriptors*, by defining a small area, called a *patch*, around a keypoint. Within the image patch, the algorithm uses a pattern template to take regular samples of intensity. It then compares preselected pairs of samples and converts them into binary strings called *feature vectors* (Figure 5-7).

A *vector* is a series of numbers. A *matrix* is a rectangular array of numbers in rows and columns that's treated as a single entity and manipulated according to rules. A *feature vector* is a matrix with one row and multiple columns. To build one, the algorithm converts the sample pairs into a binary series by concatenating a 1 to the end of the vector if the first sample has the largest intensity and a 0 if the reverse is true.

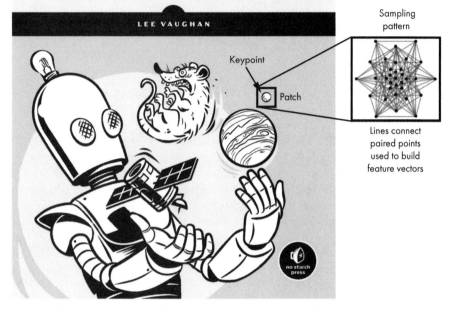

Figure 5-7: Cartoon example of how ORB generates keypoint descriptors

Some example feature vectors are shown next. I've shortened the list of vectors, because ORB usually compares and records 512 pairs of samples!

```
V₁ = [010010110100101101100--snip--]
V₂ = [100111100110010101101--snip--]
V₃ = [001101100011011101001--snip--]
--snip--
```

These descriptors act as digital fingerprints for features. OpenCV uses additional code to compensate for rotation and scale changes. This allows it to match similar features even if the feature sizes and orientations are different (see Figure 5-8).

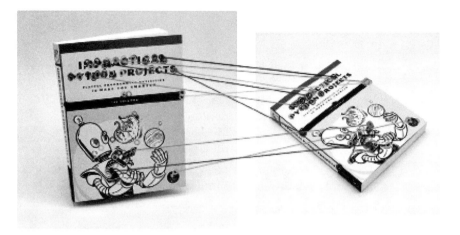

Figure 5-8: OpenCV can match keypoints despite differences in scale and orientation.

When you create the ORB object, you can specify the number of keypoints to examine. The method defaults to 500, but 100 will be more than enough for the image registration needed in this project.

Next, using the orb.detectAndCompute() method ❶, find the keypoints and their descriptors. Pass it img1 and then repeat the code for img2.

With the keypoints located and described, the next step is to find the keypoints common to both images. Start this process by creating a BFMatcher object that includes a distance measurement. The brute-force matcher takes the descriptor of one feature in the first image and compares it to all the features in the second image using the Hamming distance. It returns the closest feature.

For two strings of equal length, the *Hamming distance* is the number of positions, or indexes, at which the corresponding values are different. For the following feature vectors, the positions that don't match are shown in bold, and the Hamming distance is 3:

1001011001010
1100111001010

The bf variable will be a BFMatcher object. Call the match() method and pass it the descriptors for the two images ❷. Assign the returned list of DMatch objects to a variable named matches.

The best matches will have the lowest Hamming distance, so sort the objects in ascending order to move these to the start of the list. Note that you use a lambda function along with the object's distance attribute. A *lambda function* is a small, one-off, unnamed function defined on the fly. Words and characters that directly follow lambda are parameters. Expressions come after the colon, and returns are automatic.

Since you only need the minimum number of keypoint matches defined at the start of the program, create a new list by slicing the matches list. The best matches are at the start, so slice from the start of matches up to the value specified in MIN_NUM_KEYPOINT_MATCHES.

At this point, you're still dealing with arcane objects, as shown here:

```
best matches = [<DMatch 0000028BEBAFBFB0>, <DMatch 0000028BEBB21090>, --snip--
```

Fortunately, OpenCV knows how to handle these. Complete the function by returning the two sets of keypoints and the list of best matching objects.

Checking the Best Matches

Listing 5-5 defines a short function to let you visually check the keypoint matches. You saw the results of this function in Figure 5-5. By encapsulating these tasks in a function, you can reduce the clutter in main() and allow the user to turn off the functionality by commenting out a single line.

*blink
_comparator.py,
part 5*

```python
def QC_best_matches(img_match):
    """Draw best keypoint matches connected by colored lines."""
    cv.imshow('Best {} Matches'.format(MIN_NUM_KEYPOINT_MATCHES), img_match)
    cv.waitKey(2500)  # Keeps window active 2.5 seconds.
```

Listing 5-5: Defining a function to check the best keypoint matches

Define the function with one parameter: the matched image. This image was generated by the main() function in Listing 5-3. It consists of the left and right images with the keypoints drawn as colored circles and with colored lines connecting corresponding keypoints.

Next, call OpenCV's imshow() method to display the window. You can use the format() method when naming the window. Pass it the constant for the number of minimum keypoint matches.

Complete the function by giving the user 2.5 seconds to view the window. Note that the waitKey() method doesn't destroy the window; it just suspends the program for the allocated amount of time. After the wait period, new windows will appear as the program resumes.

Registering Images

Listing 5-6 defines the function to register the first image to the second image.

*blink
_comparator.py,
part 6*

```python
def register_image(img1, img2, kp1, kp2, best_matches):
    """Return first image registered to second image."""
    if len(best_matches) >= MIN_NUM_KEYPOINT_MATCHES:
        src_pts = np.zeros((len(best_matches), 2), dtype=np.float32)
        dst_pts = np.zeros((len(best_matches), 2), dtype=np.float32)
      ❶ for i, match in enumerate(best_matches):
            src_pts[i, :] = kp1[match.queryIdx].pt
            dst_pts[i, :] = kp2[match.trainIdx].pt
        h_array, mask = cv.findHomography(src_pts, dst_pts, cv.RANSAC)
```

```
❷ height, width = img2.shape  # Get dimensions of image 2.
  img1_warped = cv.warpPerspective(img1, h_array, (width, height))

  return img1_warped

else:
    print("WARNING: Number of keypoint matches < {}\n".format
        (MIN_NUM_KEYPOINT_MATCHES))
    return img1
```

Listing 5-6: Defining a function to register one image to another

Define a function that takes the two input images, their keypoint lists, and the list of DMatch objects returned from the find_best_matches() function as arguments. Next, load the location of the best matches into NumPy arrays. Start with a conditional to check that the list of best matches equals or exceeds the MIN_NUM_KEYPOINT_MATCHES constant. If it does, then initialize two NumPy arrays with as many rows as there are best matches.

The np.zeros() NumPy method returns a new array of a given shape and data type, filled with zeros. For example, the following snippet produces a zero-filled array three rows tall and two columns wide:

```
>>> import numpy as np
>>> ndarray = np.zeros((3, 2), dtype=np.float32)
>>> ndarray
array([[0., 0.],
       [0., 0.],
       [0., 0.]], dtype=float32)
```

In the actual code, the arrays will be at least 50×2, since you stipulated a minimum of 50 matches.

Now, enumerate the matches list and start populating the arrays with actual data ❶. For the source points, use the queryIdx.pt attribute to get the index of the descriptor in the list of descriptors for kp1. Repeat this for the next set of points, but use the trainIdx.pt attribute. The query/train terminology is a bit confusing but basically refers to the first and second images, respectively.

The next step is to apply *homography*. Homography is a transformation, using a 3×3 matrix, that maps points in one image to corresponding points in another image. Two images can be related by a homography if both are viewing the same plane from a different angle or if both images are taken from the same camera rotated around its optical axis with no shift. To run correctly, homography needs at least four corresponding points in two images.

Homography assumes that the matching points really are corresponding points. But if you look carefully at Figures 5-5 and 5-8, you'll see that the feature matching isn't perfect. In Figure 5-8, around 30 percent of the matches are incorrect!

Fortunately, OpenCV includes a findHomography() method with an outlier detector called *random sample consensus* (RANSAC). RANSAC takes random samples of the matching points, finds a mathematical model that explains their distribution, and favors the model that predicts the most points. It then discards outliers. For example, consider the points in the "Raw data" box in Figure 5-9.

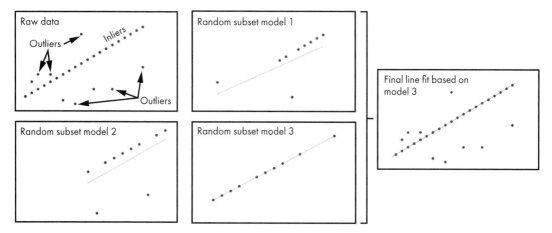

Figure 5-9: Example line fit using RANSAC to ignore outliers

As you can see, you want to fit a line through the true data points (called the *inliers*) and ignore the smaller number of spurious points (the *outliers*). Using RANSAC, you randomly sample a subset of the raw data points, fit a line to these, and then repeat this process a set number of times. Each line-fit equation would then be applied to all the points. The line that passes through the most points is used for the final line fit. In Figure 5-9, this would be the line in the rightmost box.

To run findHomography(), pass it the source and destination points and call the RANSAC method. This returns a NumPy array and a mask. The mask specifies the inlier and outlier points or the good matches and bad matches, respectively. You can use it to do tasks like draw only the good matches.

The final step is to warp the first image so that it perfectly aligns with the second. You'll need the dimensions of the second image, so use shape() to get the height and width of img2 ❷. Pass this information, along with img1 and the homography h_array, to the warpPerspective() method. Return the registered image, which will be a NumPy array.

If the number of keypoint matches is less than the minimum number you stipulated at the start of the program, the image *may not* be properly aligned. So, print a warning and return the original, nonregistered image. This will allow the main() function to continue looping through the folder images uninterrupted. If the registration is poor, the user will be aware something is wrong as the problem pair of images won't be properly aligned in the blink comparator window. An error message will also appear in the shell.

```
Comparing 2_dim_transient_left.png to 2_dim_transient_right.png.
WARNING: Number of keypoint matches < 50
```

Building the Blink Comparator

Listing 5-7 defines a function to run the blink comparator and then calls
main() if the program is run in stand-alone mode. The blink() function
loops through a specified range, showing first the registered image and
then the second image, both in the same window. It shows each image for
only one-third of a second, Clyde Tombaugh's preferred frequency when
using a blink comparator.

*blink
_comparator.py,
part 7*

```python
def blink(image_1, image_2, window_name, num_loops):
    """Replicate blink comparator with two images."""
    for _ in range(num_loops):
        cv.imshow(window_name, image_1)
        cv.waitKey(330)
        cv.imshow(window_name, image_2)
        cv.waitKey(330)

if __name__ == '__main__':
    main()
```

Listing 5-7: Defining a function to blink images on and off

Define the blink() function with four parameters: two image files, a
window name, and the number of blinks to perform. Start a for loop with
a range set to the number of blinks. Since you don't need access to the
running index, use a single underscore (_) to indicate the use of an insignificant variable. As mentioned previously in this chapter, this will prevent
code-checking programs from raising an "unused variable" warning.

Now call OpenCV's imshow() method and pass it the window name and
the first image. This will be the *registered* first image. Then pause the program for 330 milliseconds, the amount of time recommended by Clyde
Tombaugh himself.

Repeat the previous two lines of code for the second image. Because the
two images are aligned, the only thing that will change in the window are transients. If only one image contains a transient, it will appear to blink on and off.
If both images capture the transient, it will appear to dance back and forth.

End the program with the standard code that lets it run in stand-alone
mode or be imported as a module.

Using the Blink Comparator

Before you run *blink_comparator.py*, dim your room lights to simulate looking through the device's eyepieces. Then launch the program. You should
first see two obvious bright dots flashing near the center of the image. In
the next pair of images, the same dots will become very small—only a pixel
across—but you should still be able to detect them.

The third loop will show the same small transient, only this time the second image will be brighter overall than the first. You should still be able to find the transient, but it will be much more difficult. This is why Tombaugh had to carefully take and develop the images to a consistent exposure.

The fourth loop contains a single transient, shown in the left image. It should blink on and off rather than dance back and forth as in the previous images.

The fifth image pair represents control images with no transients. This is what the astronomer would see almost all the time: disappointing static star fields.

The final loop uses negative versions of the first image pair. The bright transient appears as flashing black dots. This is the type of image Clyde Tombaugh used, as it saved time. Since a black dot is as easy to spot as a white one, he felt no need to print positive images for each negative.

If you look along the left side of the registered negative image, you'll see a black stripe that represents the amount of translation needed to align the images (Figure 5-10). You won't notice this on the positive images because it blends in with the black background.

Figure 5-10: The negative image, 6_bright_transient_neg_left_registered.png

In all the loops, you may notice a dim star blinking in the upper-left corner of each image pair. This is not a transient but a false positive caused by an *edge artifact*. An edge artifact is a change to an image caused by image misalignment. An experienced astronomer would ignore this dim star because: it occurs very close to the edge of the image, and the possible transient doesn't move between images but just dims.

You can see the cause of this false positive in Figure 5-11. Because only part of a star is captured in the first frame, its brightness is reduced relative to the same star in the second image.

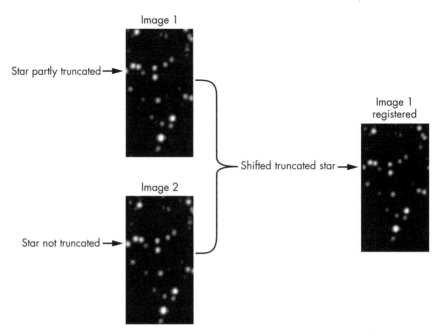

Figure 5-11: Registering a truncated star in Image 1 results in a noticeably dimmer star than in Image 2

Humans can handle edge effects intuitively, but computers require explicit rules. In the next project, you'll address this issue by excluding the edges of images when searching for transients.

Project #8: Detecting Astronomical Transients with Image Differencing

Blink comparators, once considered as important as telescopes, now sit idly gathering dust in museums. Astronomers no longer need them, as modern image-differencing techniques are much better at detecting moving objects than human eyes. Today, every part of Clyde Tombaugh's work would be done by computers.

In this project, let's pretend you're a summer intern at an observatory. Your job is to produce a digital workflow for an ancient astronomer still clinging to his rusty blink comparator.

THE OBJECTIVE

Write a Python program that takes two registered images and highlights any differences between them.

The Strategy

Instead of an algorithm that blinks the images, you now want one that automatically finds the transients. This process will still require registered images, but for convenience, just use the ones already produced in Project 7.

Detecting differences between images is a common enough practice that OpenCV ships with an absolute difference method, absdiff(), dedicated to this purpose. It takes the per-element difference between two arrays. But just detecting the differences isn't enough. Your program will need to recognize that a difference exists and show the user only the images containing transients. After all, astronomers have more important things to do, like demoting planets!

Because the objects you're looking for rest on a black background and matching bright objects are removed, any bright object remaining after differencing is worth noting. And since the odds of having more than one transient in a star field are astronomically low, flagging one or two differences should be enough to get an astronomer's attention.

The Transient Detector Code

The following *transient_detector.py* code will automate the process of detecting transients in astronomical images. Find it in the *Chapter_5* folder from the website. To avoid duplicating code, the program uses the images already registered by *blink_comparator.py*, so you'll need the *night_1_registered_transients* and *night_2* folders in the directory for this project (see Figure 5-3). As in the previous project, keep the Python code in the folder *above* these two folders.

Importing Modules and Assigning a Constant

Listing 5-8 imports the modules needed to run the program and assigns a pad constant to manage edge artifacts (see Figure 5-11). The pad represents a small distance, measured perpendicular to the image's edges, that you want to exclude from the analysis. Any objects detected between the edge of the image and the pad will be ignored.

transient _detector.py, part 1

```
import os
from pathlib import Path
import cv2 as cv

PAD = 5  # Ignore pixels this distance from edge
```

Listing 5-8: Importing modules and assigning a constant to manage edge effects

You'll need all the modules used in the previous project except for NumPy, so import them here. Set the pad distance to 5 pixels. This value may change slightly with different datasets. Later, you'll draw a rectangle around the edge space within the image so you can see how much area this parameter is excluding.

Detecting and Circling Transients

Listing 5-9 defines a function you'll use to find and circle up to two transients in each image pair. It will ignore transients in the padded area.

*transient
_detector.py,
part 2*

```
def find_transient(image, diff_image, pad):
    """Find and circle transients moving against a star field. """
    transient = False
    height, width = diff_image.shape
    cv.rectangle(image, (PAD, PAD), (width - PAD, height - PAD), 255, 1)
    minVal, maxVal, minLoc, maxLoc = cv.minMaxLoc(diff_image)
  ❶ if pad < maxLoc[0] < width - pad and pad < maxLoc[1] < height - pad:
        cv.circle(image, maxLoc, 10, 255, 0)
        transient = True
    return transient, maxLoc
```

Listing 5-9: Defining a function to detect and circle transients

The find_transient() function has three parameters: the input image, an image representing the difference between the first and second input images (representing the *difference map*), and the PAD constant. The function will find the location of the brightest pixel in the difference map, draw a circle around it, and return the location along with a Boolean indicating that an object was found.

Begin the function by setting a variable, named transient, to False. You'll use this variable to indicate whether a transient has been discovered. As transients are rare in real life, its base state should be False.

To apply the PAD constant and exclude the area near the edge of the image, you'll need the limits of the image. Get these with the shape attribute, which returns a tuple of the image's height and width.

Use the height and width variables and the PAD constant to draw a white rectangle on the image variable using OpenCV's rectangle() method. Later, this will show the user which parts of the image were ignored.

The diff_image variable is a NumPy array representing pixels. The background is black, and any "stars" that changed position (or appeared out of nowhere) between the two input images will be gray or white (see Figure 5-12).

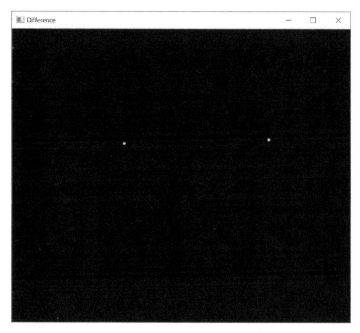

Figure 5-12: Difference image derived from the "bright transient" input images

To locate the brightest transient present, use OpenCV's `minMaxLoc()` method, which returns the minimum and maximum pixel values in the image, along with their location tuple. Note that I'm naming the variables to be consistent with OpenCV's mixed-case naming scheme (evident in names such as `maxLoc`). If you want to use something more acceptable to Python's PEP8 style guide (*https://www.python.org/dev/peps/pep-0008/*), feel free to use names like `max_loc` in place of `maxLoc`.

You may have found a maximum value near the edge of the image, so run a conditional to exclude this case by ignoring values found in the area delimited by the `PAD` constant ❶. If the location passes, circle it on the `image` variable. Use a white circle with a radius of 10 pixels and a line width of 0.

If you've drawn a circle, then you've found a transient, so set the `transient` variable to `True`. This will trigger additional activity later in the program.

End the function by returning the `transient` and `maxLoc` variables.

NOTE *The `minMaxLoc()` method is susceptible to noise, such as false positives, as it works on individual pixels. Normally, you would first run a preprocessing step, like blurring, to remove spurious pixels. This can cause you to miss dim astronomical objects, however, which can be indistinguishable from noise in a single image.*

Preparing Files and Folders

Listing 5-10 defines the `main()` function, creates lists of the filenames in the input folders, and assigns the folder paths to variables.

```
def main():
    night1_files = sorted(os.listdir('night_1_registered_transients'))
    night2_files = sorted(os.listdir('night_2'))
    path1 = Path.cwd() / 'night_1_registered_transients'
    path2 = Path.cwd() / 'night_2'
    path3 = Path.cwd() / 'night_1_2_transients'
```

Listing 5-10: Defining main()*, listing the folder contents, and assigning path variables*

Define the main() function. Then, just as you did in Listing 5-2 on page 100, list the contents of the folders containing the input images and assign their paths to variables. You'll use an existing folder to hold images containing identified transients.

Looping Through Images and Calculating Absolute Difference

Listing 5-11 starts the for loop through the image pairs. The function reads corresponding image pairs as grayscale arrays, calculates the difference between the images, and shows the result in a window. It then calls the find_transient() function on the difference image.

```
    for i, _ in enumerate(night1_files[:-1]):  # Leave off negative image
        img1 = cv.imread(str(path1 / night1_files[i]), cv.IMREAD_GRAYSCALE)
        img2 = cv.imread(str(path2 / night2_files[i]), cv.IMREAD_GRAYSCALE)

        diff_imgs1_2 = cv.absdiff(img1, img2)
        cv.imshow('Difference', diff_imgs1_2)
        cv.waitKey(2000)

        temp = diff_imgs1_2.copy()
        transient1, transient_loc1 = find_transient(img1, temp, PAD)
        cv.circle(temp, transient_loc1, 10, 0, -1)

        transient2, _ = find_transient(img1, temp, PAD)
```

Listing 5-11: Looping through the images and finding the transients

Start a for loop that iterates through the images in the *night1_files* list. The program is designed to work on *positive* images, so use image slicing ([:-1]) to exclude the negative image. Use enumerate() to get a counter; name it i, rather than _, since you'll use it as an index later.

To find the differences between images, just call the cv.absdiff() method and pass it the variables for the two images. Show the results for two seconds before continuing the program.

Since you're going to blank out the brightest transient, first make a copy of diff_imgs1_2. Name this copy temp, for temporary. Now, call the find_transient() function you wrote earlier. Pass it the first input image, the difference image, and the PAD constant. Use the results to update the transient variable and to create a new variable, transient_loc1, that records the location of the brightest pixel in the difference image.

The transient may or may not have been captured in both images taken on successive nights. To see if it was, obliterate the bright spot you just found by covering it with a black circle. Do this on the `temp` image by using black as the color and a line width of –1, which tells OpenCV to fill the circle. Continue to use a radius of 10, though you can reduce this if you're concerned the two transients will be very close together.

Call the `find_transient()` function again but use a single underscore for the location variable, as you won't be using it again. It's unlikely there'll be more than two transients present, and finding even one will be enough to open the images up to further scrutiny, so don't bother looking for more.

Revealing the Transient and Saving the Image

Listing 5-12, still in the `for` loop of the `main()` function, displays the first input image with any transients circled, posts the names of the image files involved, and saves the image with a new filename. You'll also print a log of the results for each image pair in the interpreter window.

*transient
_detector.py,
part 5*

```
if transient1 or transient2:
    print('\nTRANSIENT DETECTED between {} and {}\n'
          .format(night1_files[i], night2_files[i]))
 ❶  font = cv.FONT_HERSHEY_COMPLEX_SMALL
    cv.putText(img1, night1_files[i], (10, 25),
               font, 1, (255, 255, 255), 1, cv.LINE_AA)
    cv.putText(img1, night2_files[i], (10, 55),
               font, 1, (255, 255, 255), 1, cv.LINE_AA)

    blended = cv.addWeighted(img1, 1, diff_imgs1_2, 1, 0)
    cv.imshow('Surveyed', blended)
    cv.waitKey(2500)

 ❷  out_filename = '{}_DECTECTED.png'.format(night1_files[i][:-4])
    cv.imwrite(str(path3 / out_filename), blended)  # Will overwrite!

else:
    print('\nNo transient detected between {} and {}\n'
          .format(night1_files[i], night2_files[i]))

if __name__ == '__main__':
    main()
```

Listing 5-12: Showing the circled transients, logging the results, and saving the results

Start a conditional that checks whether a transient was found. If this evaluates to `True`, print a message in the shell. For the four images evaluated by the `for` loop, you should get this result:

```
TRANSIENT DETECTED between 1_bright_transient_left_registered.png and 1_bright_transient_right.png

TRANSIENT DETECTED between 2_dim_transient_left_registered.png and 2_dim_transient_right.png
```

```
TRANSIENT DETECTED between 3_diff_exposures_left_registered.png and 3_diff_exposures_right.png

TRANSIENT DETECTED between 4_single_transient_left_registered.png and 4_single_transient_right.png

No transient detected between 5_no_transient_left_registered.png and 5_no_transient_right.png
```

Posting a negative outcome shows that the program is working as expected and leaves no doubt that the images were compared.

Next, post the names of the two images with a positive response on the img1 array. Start by assigning a font variable for OpenCV ❶. For a listing of available fonts, search for *HersheyFonts* at *https://docs.opencv.org/4.3.0/*.

Now call OpenCV's putText() method and pass it the first input image, the filename of the image, a position, the font variable, a size, a color (white), a thickness, and a line type. The LINE_AA attribute creates an anti-aliased line. Repeat this code for the second image.

If you found two transients, you can show them both on the same image using OpenCV's addWeighted() method. This method calculates the weighted sum of two arrays. The arguments are the first image and a weight, the second image and a weight, and a scalar that's added to each sum. Use the first input image and the difference image, set the weights to 1 so that each image is used fully, and set the scalar to 0. Assign the result to a variable named blended.

Show the blended image in a window named Surveyed. Figure 5-13 shows an example outcome for the "bright" transient.

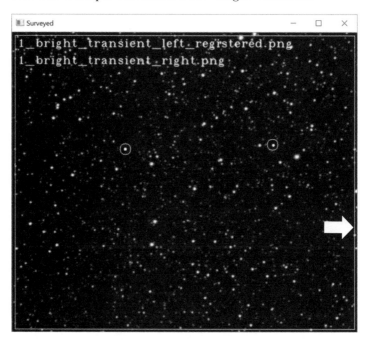

Figure 5-13: Example output window of transient_detector.py *with the pad rectangle indicated by the arrow*

Note the white rectangle near the edges of the image. This represents the PAD distance. Any transients outside this rectangle were ignored by the program.

Save the blended image using the filename of the current input image plus "DETECTED" ❷. The dim transient in Figure 5-13 would be saved as *1_bright_transient_left_registered_DECTECTED.png*. Write it to the *night_1_2_transients* folder, using the path3 variable.

If no transients were found, document the result in the shell window. Then end the program with the code to run it as a module or in stand-alone mode.

Using the Transient Detector

Imagine how happy Clyde Tombaugh would've been with your transient detector. It's truly set-it-and-forget-it. Even the changing brightness between the third pair of images, so problematic with the blink comparator, is no challenge for this program.

Summary

In this chapter, you replicated an old-time blink comparator device and then updated the process using modern computer vision techniques. Along the way, you used the pathlib module to simplify working with directory paths, and you used a single underscore for insignificant, unused variable names. You also used OpenCV to find, describe, and match interesting features in images, align the features with homography, blend the images together, and write the result to a file.

Further Reading

Out of the Darkness: The Planet Pluto (Stackpole Books, 2017), by Clyde Tombaugh and Patrick Moore, is the standard reference on the discovery of Pluto, told in the discoverer's own words.

Chasing New Horizons: Inside the Epic First Mission to Pluto (Picador, 2018), by Alan Stern and David Grinspoon, records the monumental effort to finally send a spacecraft—which, incidentally, contained Clyde Tombaugh's ashes—to Pluto.

Practice Project: Plotting the Orbital Path

Edit the *transient_detector.py* program so that if the transient is present in both input image pairs, OpenCV draws a line connecting the two transients. This will reveal the transient's orbital path against the background stars.

This kind of information was key to the discovery of Pluto. Clyde Tombaugh used the distance Pluto traveled in the two discovery plates, along with the time between exposures, to verify that the planet was near Lowell's predicted path and not just some asteroid orbiting closer to Earth.

You can find a solution, *practice_orbital_path.py*, in the appendix and in the *Chapter_5* folder.

Practice Project: What's the Difference?

The feature matching you did in this chapter has broad-reaching applications beyond astronomy. For example, marine biologists use similar techniques to identify whale sharks by their spots. This improves the accuracy of the scientists' population counts.

In Figure 5-14, something has changed between the left and right photos. Can you spot it? Even better, can you write Python programs that align and compare the two images and circle the change?

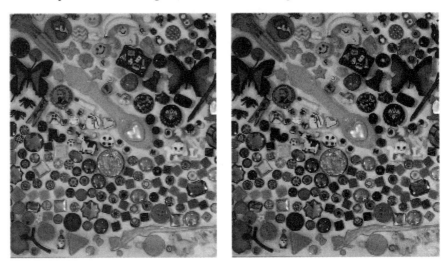

Figure 5-14: Spot the difference between the left and right images.

The starting images can be found in the *montages* folder in the *Chapter_5* folder, downloadable from the book's website. These are color images that you'll need to convert to grayscale and align prior to object detection. You can find solutions, *practice_montage_aligner.py* and *practice_montage_difference _finder.py*, in the appendix and in the *montages* folder.

Challenge Project: Counting Stars

According to *Sky and Telescope* magazine, there are 9,096 stars visible to the naked eye from both hemispheres (*https://www.skyandtelescope.com/astronomy-resources/how-many-stars-night-sky-09172014/*). That's a lot on its own, but if you look through a telescope, the number increases exponentially.

To estimate large numbers of stars, astronomers survey small regions of the sky, use a computer program to count the stars, and then extrapolate the results to larger areas. For this challenge project, pretend you're an assistant at Lowell Observatory and you're on a survey team. Write a Python program that counts the number of stars in the image *5_no_transient_left.png*, used in Projects 7 and 8.

For hints, search online for *how to count dots in an image with Python and OpenCV*. For a solution using Python and SciPy, see *http://prancer.physics .louisville.edu/astrowiki/index.php/Image_processing_with_Python_and_SciPy*. You may find your results improve if you divide the image into smaller parts.

6

WINNING THE MOON
RACE WITH APOLLO 8

In the summer of 1968, America was losing the space race. The Soviet Zond spacecraft appeared moon-ready, the Central Intelligence Agency had photographed a giant Soviet N-1 rocket sitting on its launch pad, and the Americans' troubled Apollo program still needed three more test flights. But in August, NASA manager George Low had an audacious idea. Let's go to the moon *now*. Instead of more tests in the earth's orbit, let's circle the moon in December and let *that* be the test. In that moment, the space race was essentially over. Less than a year later, the Soviets had capitulated, and Neil Armstrong had taken his great leap for all mankind.

The decision to take the Apollo 8 spacecraft to the moon was hardly trivial. In 1967, three men had died in the Apollo 1 capsule, and multiple unmanned missions had blown up or otherwise failed. Against this backdrop and with so much at stake, everything hinged on the concept of the *free return*. The mission was designed so that if the service module engine failed to fire, the ship would simply swing around the moon and return to the earth like a boomerang (Figure 6-1).

Figure 6-1: The Apollo 8 insignia, with the circumlunar free return trajectory serving as the mission number

In this chapter, you'll write a Python program that uses a drawing board module called turtle to simulate Apollo 8's free return trajectory. You'll also work with one of the classic conundrums in physics: the three-body problem.

Understanding the Apollo 8 Mission

The goal of the Apollo 8 mission was merely to circle the moon, so there was no need to take a lunar lander component. The astronauts traveled in the command and service modules, collectively known as the *CSM* (Figure 6-2).

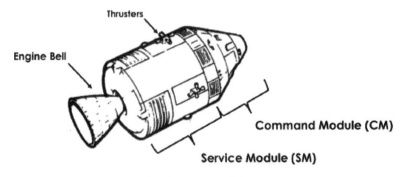

Figure 6-2: Apollo command and service modules

In the fall of 1968, the CSM engine had been tested in the earth's orbit only, and there were legitimate concerns about its reliability. To orbit the moon, the engine would have to fire twice, once to slow the spacecraft to enter lunar orbit and then again to leave orbit. With the free return trajectory, if the first maneuver failed, the astronauts could still coast home. As it turned out, the engine fired perfectly both times, and Apollo 8 orbited the moon 10 times. (The ill-fated Apollo 13, however, made great use of its free return trajectory!)

The Free Return Trajectory

Plotting a free return trajectory requires a lot of intense mathematics. It *is* rocket science, after all! Fortunately, you can simulate the trajectory in a two-dimensional graph with a few simplified parameters (Figure 6-3).

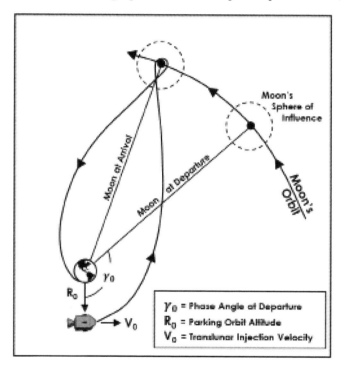

Figure 6-3: The free return trajectory (not to scale)

This 2D simulation of the free return uses a few key values: the starting position of the CSM (R_0), the velocity and orientation of the CSM (V_0), and the phase angle between the CSM and the moon (γ_0). The *phase angle*, also called the *lead angle*, is the change in the orbital time position of the CSM required to get from a starting position to a final position. The *translunar injection velocity* (V_0) is a propulsive maneuver used to set the CSM on a trajectory to the moon. It's achieved from a *parking orbit* around the earth, where the spacecraft performs internal checks and waits until the phase angle with the moon is optimal. At this point, the third stage of the *Saturn V* rocket fires and falls away, leaving the CSM to coast to the moon.

Because the moon is moving, before you perform the translunar injection, you have to predict its future position, or *lead* it, like when you're shooting skeet with a shotgun. This requires knowing the phase angle (γ_0) at the time of translunar injection. Leading the moon is a little different from shooting a shotgun, however, as space is curved and you need to factor in the gravity of the earth and the moon. The tug of these two bodies on the spacecraft creates perturbations that are difficult to calculate—so difficult, in fact, that the calculation has earned its own special name in the field of physics: the three-body problem.

The Three-Body Problem

The *three-body problem* is the challenge of predicting the behavior of three interacting bodies. Isaac Newton's gravity equations work great for predicting the behavior of two orbiting bodies, such as the earth and the moon, but add one more body to the mix, whether a spacecraft, comet, moon, or so on, and things get complicated. Newton was never able to encapsulate the behavior of three or more bodies into a simple equation. For 275 years—even with kings offering prizes for a solution—the world's greatest mathematicians worked the problem in vain.

The issue is that the three-body problem can't be solved using simple algebraic expressions or integrals. Calculating the impact of multiple gravitational fields requires numerical iteration on a scale that's impractical without a high-speed computer, such as your laptop.

In 1961, Michael Minovitch, a summer intern at the Jet Propulsion Laboratory, found the first numerical solution using an IBM 7090 mainframe, at the time the fastest computer in the world. He discovered that mathematicians could reduce the number of computations needed to solve a restricted three-body problem, like our earth-moon-CSM problem, by using a patched conic method.

The *patched conic method* is an analytical approximation that assumes you're working with a simple two-body problem while the spacecraft is in the earth's gravitational sphere of influence and another when you're within the moon's sphere of influence. It's a rough, "back-of-the-envelope" calculation that provides reasonable estimates of departure and arrival conditions, reducing the number of choices for initial velocity and position vectors. All that's left is to refine the flight path with repeated computer simulations.

Because researchers have already found and documented the Apollo 8 mission's patched conic solution, you won't need to calculate it. I've already adapted it to the 2D scenario you'll be doing here. You can experiment with alternative solutions later, however, by varying parameters such as R_0 and V_0 and rerunning the simulation.

Project #9: To the Moon with Apollo 8!

As a summer intern at NASA, you've been asked to create a simple simulation of the Apollo 8 free return trajectory for consumption by the press and general public. As NASA is always strapped for cash, you'll need to use open source software and complete the project quickly and cheaply.

THE OBJECTIVE

Write a Python program that graphically simulates the free return trajectory proposed for the Apollo 8 mission.

Using the turtle Module

To simulate the flight of Apollo 8, you'll need a way to draw and move images on the screen. There are a lot of third-party modules that can help you do this, but we'll keep things simple by using the preinstalled turtle module. Although originally invented to help kids learn programming, turtle can easily be adapted to more sophisticated uses.

The turtle module lets you use Python commands to move a small image, called a *turtle*, around a screen. The image can be invisible, an actual image, a custom shape, or one of the predefined shapes shown in Figure 6-4.

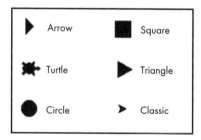

Figure 6-4: Standard turtle shapes provided with the turtle module

As the turtle moves, you can choose to draw a line behind it to trace its movement (Figure 6-5).

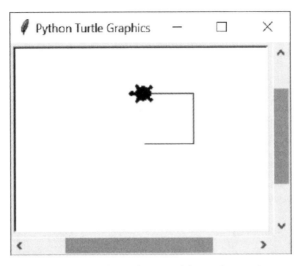

Figure 6-5: Moving the turtle around the Turtle Graphics window

This simple drawing was made with the following script:

```
>>> import turtle
>>> steve = turtle.Turtle('turtle') # Creates a turtle object with turtle shape.
>>> steve.fd(50) # Moves turtle forward 50 pixels.
>>> steve.left(90) # Rotates turtle left 90 degrees.
>>> steve.fd(50)
>>> steve.left(90)
>>> steve.fd(50)
```

You can use Python functionality with turtle to write more concise code. For example, you can use a for loop to create the same pattern.

```
>>> for i in range(3):
        steve.fd(50)
        steve.left(90)
```

Here, steve moves forward 50 pixels and then turns to the left at a right angle. These steps are repeated three times by the for loop.

Other turtle methods let you change the shape of the turtle, change its color, lift the pen so no path is drawn, "stamp" its current position on the screen, set the heading of the turtle, and get its position on the screen. Figure 6-6 shows this functionality, which is described in the script that follows.

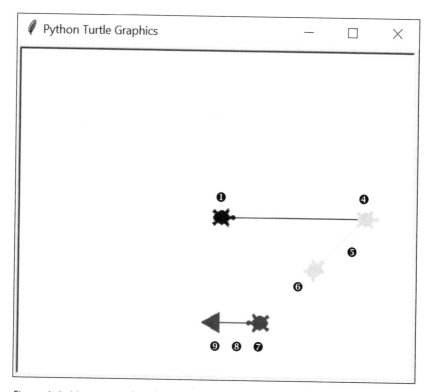

Figure 6-6: More examples of turtle behaviors. Numbers refer to script annotations.

```
>>> import turtle
>>> steve = turtle.Turtle('turtle')
❶ >>> a_stamp = steve.stamp()
❷ >>> steve.position()
❸ (0.00,0.00)
>>> steve.fd(150)
❹ >>> steve.color('gray')
>>> a_stamp = steve.stamp()
>>> steve.left(45)
❺ >>> steve.bk(75)
>>> a_stamp = steve.stamp()
❻ >>> steve.penup()
>>> steve.bk(75)
>>> steve.color('black')
❼ >>> steve.setheading(180)
>>> a_stamp = steve.stamp()
❽ >>> steve.pendown()
>>> steve.fd(50)
❾ >>> steve.shape('triangle')
```

After importing the turtle module and instantiating a turtle object named steve, leave behind an image of steve using the stamp() method ❶.

Then use the position() method ❷ to get the turtle's current (*x*, *y*) coordinates as a tuple ❸. This will come in handy when calculating the distance between objects for the gravity equation.

Move the turtle forward 150 spaces and change its color to gray ❹. Then leave a stamp behind, rotate the turtle 45 degrees, and move it backward 75 spaces using the bk() (backward) method ❺.

Leave another stamp and then stop drawing the turtle's path by using the penup() method ❻. Move steve backward another 75 spaces and color him black. Now use an alternative to rotate(), which is to directly set the heading of the turtle ❼. The heading is simply the direction the turtle is traveling. Note that the default "standard mode" directions are referenced to the east, not the north (Table 6-1).

Table 6-1: Common Directions in Degrees for the turtle Module in Standard Mode

Degrees	Direction
0	East
90	North
180	West
270	South

Leave another stamp and then put the pen down to once more draw a path behind the turtle ❽. Move steve forward 50 spaces and then change his shape to a triangle ❾. That completes the drawing.

Don't be fooled by the simplicity of what we've done so far. With the right commands, you can draw intricate designs, such as the Penrose tiling in Figure 6-7.

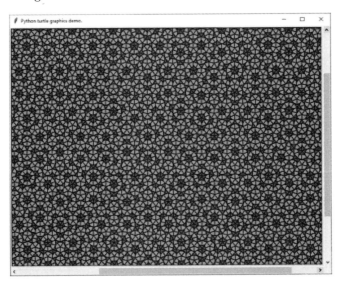

Figure 6-7: A Penrose tiling produced by the turtle *module demo,* penrose.py

The turtle module is part of the Python Standard Library, and you can find the official documentation at *https://docs.python.org/3/library/turtle.html ?highlight=turtle#module-turtle/*. For a quick tutorial, do an online search for Al Sweigart's *Simple Turtle Tutorial for Python*.

The Strategy

We've now made a strategic decision to use turtle to draw the simulation, but how should the simulation look? For convenience, I'd suggest basing it on Figure 6-3. You'll start with the CSM in the same parking orbit position around the earth (R_0) and the moon at the same approximate phase angle (γ_0). You can use images to represent the earth and the moon and custom turtle shapes to build the CSM.

Another big decision at this point is whether to use procedural or object-oriented programming (OOP). When you plan to generate multiple objects that behave similarly and interact with each other, OOP is a good choice. You can use an OOP class as a blueprint for the earth, the moon, and the CSM objects and automatically update the object attributes as the simulation runs.

You can run the simulation using *time steps*. Basically, each program loop will represent one unit of dimensionless time. With each loop, you'll need to calculate each object's position and update (redraw) it on the screen. This requires solving the three-body problem. Fortunately, not only has someone done this already, they've done it using turtle.

Python modules often include example scripts to show you how to use the product. For instance, the matplotlib gallery includes code snippets and tutorials for making a huge number of charts and plots. Likewise, the turtle module comes with *turtle-example-suite*, which includes demonstrations of turtle applications.

One of the demos, *planet_and_moon.py*, provides a nice "recipe" for handling a three-body problem in turtle (Figure 6-8). To see the demos, open a PowerShell or terminal window and enter **python -m turtledemo**. Depending on your platform and how many versions of Python you have installed, you may need to use python3 -m turtledemo.

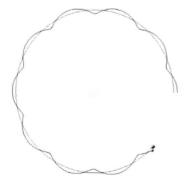

Figure 6-8: Screen capture of the planet_and_moon.py *turtle demo*

This demo addresses the sun-earth-moon three-body problem, but it can be easily adapted to handle an earth-moon-CSM problem. Again, for the specific Apollo 8 situation, you'll use Figure 6-3 to guide development of the program.

The Apollo 8 Free Return Code

The *apollo_8_free_return.py* program uses turtle graphics to generate a top-down view of the Apollo 8 CSM leaving the earth's orbit, circling the moon, and returning to the earth. The core of the program is based on the *planet_and_moon.py* demo discussed in the previous section.

You can find the program in the *Chapter_6* folder, downloadable from the book's website at *https://nostarch.com/real-world-python/*. You'll also need the earth and moon images found there (Figure 6-9). Be sure to keep them in the same folder as the code and don't rename them.

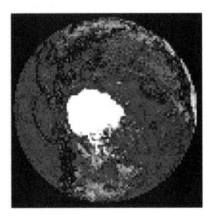

Figure 6-9: earth_100x100.gif *and* moon_27x27.gif *images used in the simulation*

Importing turtle and Assigning Constants

Listing 6-1 imports the turtle module and assigns constants that represent key parameters: the gravitational constant, the number of times to run the main loop, and the x and y values for R_0 and V_0 (see Figure 6-3). Listing these values near the top of the program makes them easy to find and alter later.

*apollo_8_free
_return.py*, part 1

```
from turtle import Shape, Screen, Turtle, Vec2D as Vec

# User input:
G = 8
NUM_LOOPS = 4100
Ro_X = 0
Ro_Y = -85
Vo_X = 485
Vo_Y = 0
```

Listing 6-1: *Importing turtle and assigning constants*

You'll need to import four helper classes from turtle. You'll use the Shape class to make a custom turtle that looks like the CSM. The Screen subclass makes the screen, called a *drawing board* in turtle parlance. The Turtle subclass creates the turtle objects. The Vec2D import is a two-dimensional vector class. It will help you define velocity as a vector of magnitude and direction.

Next, assign some variables that the user may want to tweak later. Start with the gravitational constant, used in Newton's gravity equations to ensure the units come out right. Assign it 8, the value used in the turtle demo. Think of this as a *scaled* gravitational constant. You can't use the true constant, as the simulation doesn't use real-world units.

You'll run the simulation in a loop, and each iteration will represent a time step. With each step, the program will recalculate the position of the CSM as it moves through the gravity fields of the earth and the moon. The value of 4100, arrived at by trial and error, will stop the simulation just after the spacecraft arrives back on the earth.

In 1968, a round-trip to the moon took about six days. Since you're incrementing the time unit by 0.001 with each loop and running 4,100 loops, this means a time step in the simulation represents about two minutes of time in the real world. The longer the time step, the faster the simulation but the less accurate the results, as small errors compound over time. In actual fight path simulations, you can optimize the time step by first running a small step, for maximum accuracy, and then using the results to find the largest time step that yields a similar result.

The next two variables, Ro_X and Ro_Y, represent the (*x*, *y*) coordinates of the CSM at the time of the translunar injection (see Figure 6-3). Likewise, Vo_X and Vo_Y represent the *x*- and *y*-direction components of the translunar injection velocity, which is applied by the third stage of the *Saturn V* rocket. These values started out as best guesses and were refined with repeated simulations.

Creating a Gravity System

Because the earth, the moon, and CSM form a continuously interacting gravity system, you'll want a convenient way to represent them and their respective forces. For this, you'll need two classes, one to create a gravity system and one to create the bodies within it. Listing 6-2 defines the GravSys class that helps you create a mini solar system. This class will use a list to keep track of all the bodies in motion and loop them through a series of time steps. It's based on the *planet_and_moon.py* demo in the turtle library.

*apollo_8_free
_return.py, part 2*

```
class GravSys():
    """Runs a gravity simulation on n-bodies."""

    def __init__(self):
        self.bodies = []
        self.t = 0
        self.dt = 0.001
```

```
❶ def sim_loop(self):
      """Loop bodies in a list through time steps."""
      for _ in range(NUM_LOOPS):
          self.t += self.dt
          for body in self.bodies:
              body.step()
```

Listing 6-2: Defining a class to manage the bodies in the gravity system

The GravSys class defines how long the simulation will run, how much time will pass between time steps (loops), and what bodies will be involved. It also calls the step() method of the Body class you'll define in Listing 6-3. This method will update each body's position as a result of gravitational acceleration.

Define the initialization method and, as per convention, pass it self as a parameter. The self parameter represents the GravSys object you'll create later in the main() function.

Create an empty list named bodies to hold the earth, the moon, and the CSM objects. Then assign attributes for when the simulation starts and the amount to increment time with each loop, known as *delta time* or dt. Set the starting time to 0 and set the dt time step to 0.001. As discussed in the previous section, this time step will correspond to about two minutes in the real world and will produce a smooth, accurate, and fast simulation.

The last method controls the time steps in the simulation ❶. It uses a for loop with the range set to the NUM_LOOPS variable. Use a single underscore (_) rather than i to indicate the use of an insignificant variable (see Listing 5-3 in Chapter 5 for details).

With each loop, increment the gravity system's time variable by dt. Then, apply the time shift to each body by looping through the list of bodies and calling the body.step() method, which you'll define later within the Body class. This method updates the position and velocity of the bodies due to gravitational attraction.

Creating Celestial Bodies

Listing 6-3 defines the Body class used to build the earth, the moon, and the CSM Body objects. Although no one would ever mistake a planet for a small spacecraft, they're not that different from a gravitational standpoint, and you can stamp them both out of the same mold.

*apollo_8_free
_return.py, part 3*

```
class Body(Turtle):
    """Celestial object that orbits and projects gravity field."""
    def __init__(self, mass, start_loc, vel, gravsys, shape):
        super().__init__(shape=shape)
        self.gravsys = gravsys
        self.penup()
        self.mass = mass
        self.setpos(start_loc)
        self.vel = vel
```

```
gravsys.bodies.append(self)
#self.resizemode("user")
#self.pendown()  # Uncomment to draw path behind object.
```

Listing 6-3: Defining a class to create objects for the earth, the moon, and the CSM

Define a new class by using the Turtle class as its *ancestor*. This means the Body class will conveniently inherit all the Turtle class's methods and attributes.

Next, define an initializer method for the body object. You'll use this to create new Body objects in the simulation, a process called *instantiation* in OOP. As parameters, the initialize method takes itself, a mass attribute, a starting location, a starting velocity, the gravity system object, and a shape.

The super() function lets you invoke the method of a superclass to gain access to inherited methods from the ancestor class. This allows your Body objects to use attributes from the prebuilt Turtle class. Pass it the shape attribute, which will allow you to pass a custom shape or image to your bodies when you build them in the main() function.

Next, assign an instance attribute for the gravsys object. This will allow the gravity system and body to interact. Note that it's best to initialize attributes through the __init__() method, as we do in this case, since it's the first method called after the object is created. This way, these attributes will be immediately available to any other methods in the class, and other developers can see a list of all the attributes in one place.

The following penup() method of the Turtle class will remove the drawing pen so the object doesn't leave a path behind it as it moves. This gives you the option of running the simulation with and without visible orbital paths.

Initialize a mass attribute for the body. You'll need this to calculate the force of gravity. Next, assign the body's starting position using the setpos() method of the Turtle class. The starting position of each body will be an (x, y) tuple. The origin point $(0, 0)$ will be at the center of the screen. The x-coordinate increases to the right, and the y-coordinate increases upward.

Assign an initialization attribute for velocity. This will hold the starting velocity for each object. For the CSM, this value will change throughout the simulation as the ship moves through the gravity fields of the earth and the moon.

As each body is instantiated, use dot notation to append it to the list of bodies in the gravity system. You'll create the gravsys object from the GravSys() class in the main() function.

The final two lines, commented out, allow the user to change the simulation window size and choose to draw a path behind each object. Start out with a full-screen display and keep the pen in the up position to let the simulation run quickly.

Calculating Acceleration Due to Gravity

The Apollo 8 simulation will begin immediately after the translunar injection. At this point, the third stage of the *Saturn V* has fired and fallen away, and the CSM is beginning its coast to the moon. All changes in velocity or direction will be entirely due to changes in gravitational force.

The method in Listing 6-4 loops through the bodies in the bodies list, calculates acceleration due to gravity for each body, and returns a vector representing the body's acceleration in the *x* and *y* directions.

apollo_8_free
_return.py, part 4

```
def acc(self):
    """Calculate combined force on body and return vector components."""
    a = Vec(0, 0)
    for body in self.gravsys.bodies:
        if body != self:
            r = body.pos() - self.pos()
            a += (G * body.mass / abs(r)**3) * r
    return a
```

Listing 6-4: Calculating acceleration due to gravity

Still within the Body class, define the acceleration method, called acc(), and pass it self. Within the method, name a local variable a, again for acceleration, and assign it to a vector tuple using the Vec2D helper class. A 2D vector is a pair of real numbers (a, b), which in this case represent *x* and *y* components, respectively. The Vec2D helper class enforces rules that permit easy mathematical operations using vectors, as follows:

- $(a, b) + (c, d) = (a + c, b + d)$
- $(a, b) - (c, d) = (a - c, b - d)$
- $(a, b) \times (c, d) = ac + bd$

Next, start looping through the items in the bodies list, which contains the earth, the moon, and the CSM. You'll use the gravitational force of each body to determine the acceleration of the object for which you're calling the acc() method. It doesn't make sense for a body to accelerate itself, so exclude the body if it's the same as self.

To calculate gravitational acceleration (stored in the g variable) at a point in space, you'll use the following formula:

$$g = \frac{GM}{r^2} \hat{r}$$

where *M* is the mass of the attracting body, *r* is the distance (radius) between bodies, *G* is the gravitational constant you defined earlier, and $\hat{r}$ is the unit vector from the center of mass of the attracting body to the center of mass of the body being accelerated. The *unit vector*, also known as the *direction vector* or *normalized vector*, can be described as $r/|r|$, or:

$$\frac{(position\ of\ attracting\ body\ -\ position\ of\ body\ being\ attracted)}{|(position\ of\ attracting\ body\ -\ position\ of\ body\ being\ attracted)|}$$

The unit vector allows you to capture the direction of acceleration, which will be either positive or negative. To calculate the unit vector, you'll have to calculate the distance between bodies by using the turtle pos() method to get each body's current position as a Vec2D vector. As described previously, this is a tuple of the (x, y) coordinates.

You'll then input that tuple into the acceleration equation. Each time you loop through a new body, you'll change the a variable based on the gravitational pull of the body being examined. For example, while the earth's gravity may slow the CSM, the moon's gravity may pull in the opposite direction and cause it to speed up. The a variable will capture the net effect at the end of the loop. Complete the method by returning a.

Stepping Through the Simulation

Listing 6-5, still in the Body class, defines a method to solve the three-body problem. It updates the position, orientation, and velocity of bodies in the gravity system with each time step. The shorter the time steps, the more accurate the solution, though at the cost of computational efficiency.

apollo_8_free _return.py, part 5

```
def step(self):
    """Calculate position, orientation, and velocity of a body."""
    dt = self.gravsys.dt
    a = self.acc()
    self.vel = self.vel + dt * a
    self.setpos(self.pos() + dt * self.vel)
❶   if self.gravsys.bodies.index(self) == 2:   # Index 2 = CSM.
        rotate_factor = 0.0006
        self.setheading((self.heading() - rotate_factor * self.xcor()))
❷       if self.xcor() < -20:
            self.shape('arrow')
            self.shapesize(0.5)
            self.setheading(105)
```

Listing 6-5: Applying the time step and rotating the CSM

Define a step() method to calculate position, orientation, and velocity of a body. Assign it self as an argument.

Within the method definition, set a local variable, dt, to the gravsys object of the same name. This variable has no link to any real-time system; it's just a floating-point number that you'll use to increment velocity with each time step. The larger the dt variable is, the faster the simulation will run.

Now call the self.acc() method to calculate the acceleration that the current body experiences due to the combined gravitational fields of the other bodies. This method returns a vector tuple of (x, y) coordinates. Multiply it by dt and add the results to self.vel(), which is also a vector, to update the body's velocity for the current time step. Recall that, behind the scenes, the Vec2D class will manage the vector arithmetic.

To update the body's position in the turtle graphics window, multiply the body's velocity by the time step and add the result to the body's position attribute. Now each body will move according to the gravitational pull of the other bodies. You just solved the three-body problem!

Next, add some code to refine the CSM's behavior. Thrust comes out of the back of the CSM, so in real missions, the rear of the spacecraft is oriented toward its target. This way, the engine can fire and slow the ship enough to enter lunar orbit or the earth's atmosphere. Orienting the ship this way isn't necessary with a free return trajectory, but since Apollo 8 planned to fire its engines and enter lunar orbit (and did), you should orient the ship properly throughout its journey.

Start by selecting the CSM from the list of bodies ❶. In the `main()` function, you'll create the bodies in order of size, so the CSM will be the third item in the list, at index 2.

To get the CSM to rotate as it coasts through space, assign a small number to a local variable named `rotate_factor`. I arrived at this number through trial and error. Next, set the heading of the CSM turtle object using its `selfheading` attribute. Instead of passing it (x, y) coordinates, call the `self.heading()` method, which returns the object's current heading in degrees, and subtract from it the `rotate_factor` variable multiplied by the body's current x location, obtained by calling the `self.xcor()` method. This will cause the CSM to rotate faster as it approaches the moon to keep its tail pointed in the direction of travel.

You'll need to eject the service module before the spacecraft enters the earth's atmosphere. To do this at a position similar to that in real Apollo missions, use another conditional to check the spacecraft's x-coordinate ❷. The simulation expects the earth to be near the center of the screen, at coordinates $(0, 0)$. In turtle, the x-coordinate will decrease as you move left of the center and increase as you move to the right. If the CSM's x-coordinate is less than –20 pixels, you can assume that it's returning home and that it's time to part company with the service module.

You'll model this event by changing the shape of the turtle representing the CSM. Since turtle includes a standard shape—called `arrow`—that looks similar to the command module, all you need to do now is call the `self.shape()` method and pass it the name of the shape. Then call the `self.shapesize()` method and halve the size of the arrow to make it match the command module in the CSM custom shape, which you'll make later. When the CSM passes the –20 x-position, the service module will magically disappear, leaving the command module to complete the voyage home.

Finally, you'll want to orient the base of the command module, with its heat-resistant shielding, toward the earth. Do this by setting the arrow shape's heading to 105 degrees.

Defining main(), Setting Up the Screen, and Instantiating the Gravity System

You used object-oriented programming to build the gravity system and the bodies within it. To run the simulation, you'll return to procedural programming and use a `main()` function. This function sets up the turtle graphics screen, instantiates objects for the gravity system and the three bodies, builds a custom shape for the CSM, and calls the gravity system's `sim_loop()` method to walk through the time steps.

Listing 6-6 defines `main()` and sets up the screen. It also creates a gravity system object to manage your mini solar system.

apollo_8_free_
return.py, part 6

```
def main():
    screen = Screen()
    screen.setup(width=1.0, height=1.0) # For fullscreen.
    screen.bgcolor('black')
    screen.title("Apollo 8 Free Return Simulation")

    gravsys = GravSys()
```

Listing 6-6: Setting up the screen and making a gravsys object in main()

Define `main()` and then instantiate a screen object (a drawing window) based on the `TurtleScreen` subclass. Then invoke the screen object's `setup()` method to set the size of screen to full. Do this by passing `width` and `height` arguments of 1.

If you don't want the drawing window to take up the full screen, pass `setup()` the pixel arguments shown in the following snippet:

```
screen.setup(width=800, height=900, startx=100, starty=0)
```

Note that a negative startx value uses right justification, a negative starty uses bottom alignment, and the default settings create a centered window. Feel free to experiment with these parameters to get the best fit to your monitor.

Complete setting up the screen by setting its background color to black and giving it a title. Next, instantiate a gravity system object, gravsys, using the GravSys class. This object will give you access to the attributes and methods in the GravSys class. You'll pass it to each body when you instantiate them shortly.

Creating the Earth and Moon

Listing 6-7, still in the `main()` function, creates turtle objects for the earth and the moon using the Body class you defined earlier. The earth will remain stationary at the center of the screen, while the moon will revolve around the earth.

When you create these objects, you'll set their starting coordinates. The starting position of the earth is near the center of the screen, biased downward a bit to give the moon and CSM room to interact near the top of the window.

The starting position of the moon and CSM should reflect what you see in Figure 6-3, with the CSM vertically beneath the center of the earth. This way, you only need to thrust in the *x* direction, rather than calculate a vector component velocity that includes some movement in the *x* direction and some in the *y* direction.

```
image_earth = 'earth_100x100.gif'
screen.register_shape(image_earth)
earth = Body(1000000, (0, -25), Vec(0, -2.5), gravsys, image_earth)
earth.pencolor('white')
earth.getscreen().tracer(n=0, delay=0)

❶ image_moon = 'moon_27x27.gif'
screen.register_shape(image_moon)
moon = Body(32000, (344, 42), Vec(-27, 147), gravsys, image_moon)
moon.pencolor('gray')
```

Listing 6-7: Instantiating turtles for the earth and moon

Start by assigning the image of the earth, which is included in the folder for this project, to a variable. Note that images should be *gif* files and cannot be rotated to show the turtle's heading. So that turtle recognizes the new shape, add it to the TurtleScreen shapelist using the screen.register_shape() method. Pass it the variable that references the earth image.

Now it's time to instantiate the turtle object for the earth. You call the Body class and pass it the arguments for mass, starting position, starting velocity, gravity system, and turtle shape—in this case, the image. Let's talk about each of these arguments in more detail.

You're not using real-world units here, so mass is an arbitrary number. I started with the value used for the sun in the turtle demo *planet_and_moon.py*, on which this program is based.

The starting position is an (x, y) tuple that places the earth near the center of the screen. It's biased downward 25 pixels, however, as most of the action will take place in the upper quadrant of the screen. This placement will provide a little more room in that region.

The starting velocity is a simple (x, y) tuple provided as an argument to the Vec2D helper class. As discussed previously, this will allow later methods to alter the velocity attribute using vector arithmetic. Note that the earth's velocity is not (0, 0), but (0, -2.5). In real life and in the simulation, the moon is massive enough to affect the earth so that the center of gravity between the two is not at the center of the earth, but farther out. This will cause the earth turtle to wobble and shift positions in a distracting manner during the simulation. Because the moon will be in the upper part of the screen during simulation, shifting the earth downward a small amount each time step will dampen the wobbling.

The last two arguments are the gravsys object you instantiated in the previous listing and the image variable for the earth. Passing gravsys means the earth turtle will be added to the list of bodies and included in the sim_loop() class method.

Note that if you don't want to use a lot of arguments when instantiating an object, you can change an object's attributes after it's created. For example, when defining the Body class, you could've set self.mass = 0, rather than using an argument for mass. Then, after instantiating the earth body, you could reset the mass value using earth.mass = 1000000.

Because the earth wobbles a little, its orbital path will form a tight circle at the top of the planet. To hide it in the polar cap, use the turtle pencolor() method and set the line color to white.

Finish the earth turtle with code that delays the start of the simulation and prevents the various turtles from flashing on the screen as the program first draws and resizes them. The getscreen() method returns the TurtleScreen object the turtle is drawing on. TurtleScreen methods can then be called for that object. In the same line, call the tracer() method that turns the turtle animation on or off and sets a delay for drawing updates. The n parameter determines the number of times the screen updates. A value of 0 means the screen updates with every loop; larger values progressively repress the updates. This can be used to accelerate the drawing of complex graphics, but at the cost of image quality. The second argument sets a delay value, in milliseconds, between screen updates. Increasing the delay slows the animation.

You'll build the moon turtle in a similar fashion to the one for the earth. Start by assigning a new variable to hold the moon image ❶. The moon's mass is only a few percent of the earth's mass, so use a much smaller value for the moon. I started out with a mass of around 16,000 and tweaked the value until the CSM's flight path produced a visually pleasing loop around the moon.

The moon's starting position is controlled by the phase angle shown in Figure 6-3. Like this figure, the simulation you're creating here is not to scale. Although the earth and moon images will have the correct relative sizes, the distance between the two is smaller than the actual distance, so the phase angle will need to be adjusted accordingly. I've reduced the distance in the model because space is big. Really big. If you want to show the simulation to scale and fit it all on your computer monitor, then you must settle for a ridiculously tiny earth and moon (Figure 6-10).

Figure 6-10: Earth and moon system at closest approach, or perigee, shown to scale

To keep the two bodies recognizable, you'll instead use larger, properly scaled images but reduce the distance between them (Figure 6-11). This configuration will be more relatable to the viewer and still allow you to replicate the free return trajectory.

Because the earth and the moon are closer together in the simulation, the moon's orbital velocity will be faster than in real life, as per Kepler's second law of planetary motion. To compensate for this, the moon's starting position is designed to reduce the phase angle compared to that shown in Figure 6-3.

Figure 6-11: The earth and moon system in the simulation, with only the body sizes at the correct scale

Finally, you'll want the option to draw a line behind the moon to trace its orbit. Use the turtle pencolor() method and set the line color to gray.

NOTE *Parameters such as mass, initial position, and initial velocity are good candidates for global constants. Despite this, I chose to enter them as method arguments to avoid overloading the user with too many input variables at the start of the program.*

Building a Custom Shape for the CSM

Now it's time to instantiate a turtle object to represent the CSM. This requires a little more work than the last two objects.

First, there's no way to show the CSM at the same scale as the earth and the moon. To do that, you'd need *less than* a pixel, which is impossible. Plus, where's the fun in that? So, once again, you'll take liberties with scale and make the CSM large enough to be recognizable as an Apollo spacecraft.

Second, you won't use an image for the CSM, as you did with the other two bodies. Because image shapes don't automatically rotate when a turtle turns and you want to orient the CSM tail-first through most of its journey, you must instead customize your own shape.

Listing 6-8, still in main(), builds a representation of the CSM by drawing basic shapes, such as rectangles and triangles. You then combine these individual primitives into a final compound shape.

apollo_8_free_ return.py, part 8

```
csm = Shape('compound')
cm = ((0, 30), (0, -30), (30, 0))
csm.addcomponent(cm, 'white', 'white')
sm = ((-60, 30), (0, 30), (0, -30), (-60, -30))
csm.addcomponent(sm, 'white', 'black')
nozzle = ((-55, 0), (-90, 20), (-90, -20))
csm.addcomponent(nozzle, 'white', 'white')
screen.register_shape('csm', csm)
```

Listing 6-8: Building a custom shape for the CSM turtle

Name a variable csm and call the turtle Shape class. Pass it 'compound', indicating you want to build the shape using multiple components.

The first component will be the command module. Name a variable cm and assign it to a tuple of coordinate pairs, known as a *polygon type* in turtle. These coordinates build a triangle, as shown in Figure 6-12.

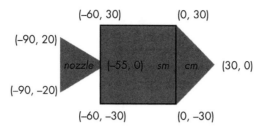

Figure 6-12: CSM compound shape with coordinates for nozzle, service module, and command module

Add this triangle component to the csm shape using the addcomponent() method, called with dot notation. Pass it the cm variable, a fill color, and an outline color. Good fill colors are white, silver, gray, or red.

Repeat this general process for the service module rectangle. Set the outline color to black when you add the component to delineate the service and command modules (see Figure 6-12).

Use another triangle for the nozzle, also called the *engine bell*. Add the component and then register the new csm compound shape to the screen. Pass the method a name for the shape and then the variable referencing the shape.

Creating the CSM, Starting the Simulation, and Calling main()

Listing 6-9 completes the main() function by instantiating a turtle for the CSM and calling the simulation loop that runs the time steps. It then calls main() if the program is run in stand-alone mode.

*apollo_8_free
_return.py, part 9*

```
ship = Body(1, (Ro_X, Ro_Y), Vec(Vo_X, Vo_Y), gravsys, 'csm')
ship.shapesize(0.2)
ship.color('white')
ship.getscreen().tracer(1, 0)
ship.setheading(90)

gravsys.sim_loop()

if __name__ == '__main__':
    main()
```

Listing 6-9: Instantiating a CSM turtle, calling the simulation loop and main()

Create a turtle named ship to represent the CSM. The starting position is an (*x*, *y*) tuple that places the CSM in a parking orbit directly below the earth on the screen. I first approximated the proper height for the parking orbit (R_0 in Figure 6-3) and then fine-tuned it by repeatedly running the simulation. Note that you use the constants assigned at the start of the program, rather than actual values. This is to make it easier for you to experiment with these values later.

The velocity argument (Vo_X, Vo_Y) represents the speed of the CSM at the moment the Saturn third stage stops firing during translunar injection. All the thrust is in the *x* direction, but the earth's gravity will cause the flight path to immediately curve upward. Like the R_0 parameter, a best-guess velocity was input and refined through simulation. Note that the velocity is a tuple input using the Vec2D helper class, which allows later methods to alter the velocity using vector arithmetic.

Next, set the size of the ship turtle using the shapesize() method. Then set its path color to white so it will match the ship color. Other attractive colors are silver, gray, and red.

Control the screen updates with the getscreen() and tracer() methods, described in Listing 6-7, and then set the ship's heading to 90 degrees, which will point it due east on the screen.

That completes the body objects. Now all that's left is to launch the simulation loop, using the gravsys object's sim_loop() method. Back in the global space, finish the program with the code to run the program as an imported module or in stand-alone mode.

As the program is currently written, you'll have to manually close the Turtle Graphics window. If you want the window to close automatically, add the following command as the last line in main():

```
screen.bye()
```

Running the Simulation

When you first run the simulation, the pen will be up, and none of the bodies will draw their orbital path (Figure 6-13). The CSM will smoothly rotate and reorient itself as it approaches the moon and then the earth.

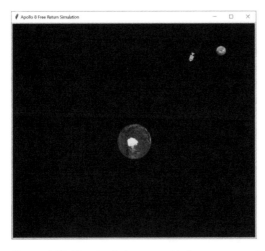

Figure 6-13: The simulation run with the pen up and the CSM approaching the moon

To trace the journey of the CSM, go to the definition of the Body class and uncomment this line:

```
self.pendown() # uncomment to draw path behind object
```

You should now see the figure-eight shape of the free return trajectory (Figure 6-14).

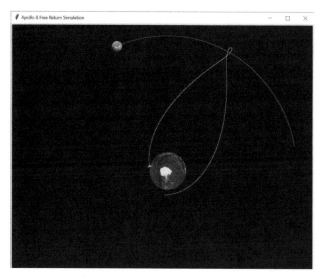

Figure 6-14: The simulation run with the pen down and the CM at splashdown in the Pacific

You can also simulate *gravity propulsion*—otherwise known as a *slingshot maneuver*—by setting the Vo_X velocity variable to a value between 520 and 540 and rerunning the simulation. This will cause the CSM to pass behind the moon and steal some of its momentum, increasing the ship's velocity and deflecting its flight path (Figure 6-15). Bye-bye Apollo 8!

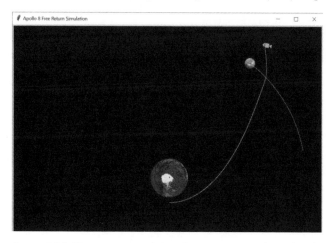

Figure 6-15: The gravitational slingshot maneuver achieved with Vo_X = 520

This project should teach you that space travel is a game of seconds and centimeters. If you continue to experiment with value of the Vo_X variable, you'll find that even small changes can doom the mission. If you don't crash into the moon, you'll reenter the earth's atmosphere too steeply or miss it entirely!

The nice thing about simulations is that, if you fail, you can live to try again. NASA runs countless simulations for all its proposed missions. The results help NASA choose between competing flight plans, find the most efficient routes, decide what to do if things go wrong, and much more.

Simulations are especially important for outer solar system exploration, where great distances make real-time communications impossible. The timing of key events, such as firing thrusters, taking photographs, or dropping probes, are all preprogrammed based on meticulous simulations.

Summary

In this chapter, you learned how to use the turtle drawing program, including how to make customized turtle shapes. You also learned how to use Python to simulate gravity and solve the famous three-body problem.

Further Reading

Apollo 8: The Thrilling Story of the First Mission to the Moon (Henry Holt and Co., 2017), by Jeffrey Kluger, covers the historic Apollo 8 mission from its unlikely beginning to its "unimaginable triumph."

An online search for *PBS Nova How Apollo 8 Left Earth Orbit* should return a short video clip on the Apollo 8 translunar injection maneuver, marking the first time humans left the earth's orbit and traveled to another celestial body.

NASA Voyager 1 & 2 Owner's Workshop Manual (Haynes, 2015), by Christopher Riley, Richard Corfield, and Philip Dolling, provides interesting background on the three-body problem and Michael Minovitch's many contributions to space travel.

The Wikipedia *Gravity assist* page contains lots of interesting animations of various gravity-assist maneuvers and historic planetary flybys that you can reproduce with your Apollo 8 simulation.

Chasing New Horizons: Inside the Epic First Mission to Pluto (Picador, 2018), by Alan Stern and David Grinspoon, documents the importance—and ubiquity—of simulations in NASA missions.

Practice Project: Simulating a Search Pattern

In Chapter 1, you used Bayes' rule to help the Coast Guard search for a sailor lost at sea. Now, use turtle to design a helicopter search pattern to find the missing sailor. Assume the spotters can see for 20 pixels and make the spacing between long tracks 40 pixels (see Figure 6-16).

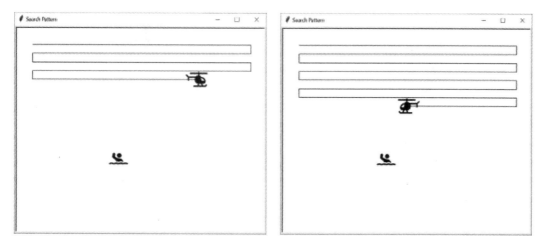

Figure 6-16: Two screenshots from practice_search_pattern.py

For fun, add a helicopter turtle and orient it properly for each pass. Also add a randomly positioned sailor turtle, stop the simulation when the sailor is found, and post the joyous news to the screen (Figure 6-17).

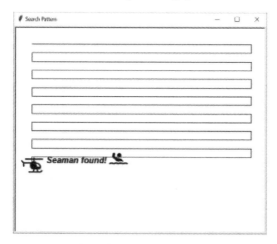

Figure 6-17: The sailor is spotted in practice_search_pattern.py.

You can find a solution, *practice_search_pattern.py*, in the appendix. I've included a digital version, along with helicopter and sailor images, in the *Chapter_6* folder, downloadable from the book's website.

Practice Project: Start Me Up!

Rewrite *apollo_8_free_return.py* so that a moving moon approaches a stationary CSM, causes the CSM to start moving, and then swings it up and away. For fun, orient the CSM turtle so that it always points in the direction of travel, as if under its own propulsion (see Figure 6-18).

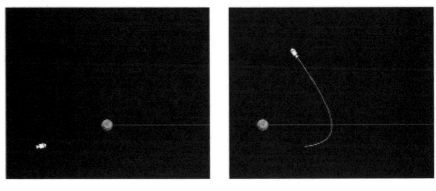

Figure 6-18: The moon approaches a stationary CSM (left) and then flings it to the stars (right).

For a solution, see *practice_grav_assist_stationary.py* in the appendix or download it from *https://nostarch.com/real-world-python/*.

Practice Project: Shut Me Down!

Rewrite *apollo_8_free_return.py* so that the CSM and moon have crossing orbits, the CSM passes before the moon, and the moon's gravity slows the CSM's progress to a crawl while changing its direction by about 90 degrees. As in the previous practice project, have the CSM point in the direction of travel (see Figure 6-19).

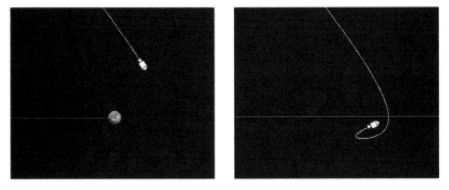

Figure 6-19: The moon and CSM cross orbits, and the moon slows and turns the CSM.

For a solution, see *practice_grav_assist_intersecting.py* in the appendix or download it from *https://nostarch.com/real-world-python/*.

Challenge Project: True-Scale Simulation

Rewrite *apollo_8_free_return.py* so that the earth, the moon, and the distance between them are all accurately scaled, as shown in Figure 6-10. Use colored circles, rather than images, for the earth and the moon and make the CSM invisible (just draw a line behind it). Use Table 6-2 to help determine the relative sizes and distances to use.

Table 6-2: Length Parameters for the Earth-Moon System

Earth radius	6,371 km
Moon radius	1,737 km
Earth-moon distance	356,700 km*

*Closest approach during Apollo 8 mission in December 1968

Challenge Project: The Real Apollo 8

Rewrite *apollo_8_free_return.py* so that it simulates the entire Apollo 8 mission, not just the free return component. The CSM should orbit the moon 10 times before returning to the earth.

7

SELECTING MARTIAN LANDING SITES

Landing a spacecraft on Mars is extraordinarily difficult and fraught with peril. No one wants to lose a billion-dollar probe, so engineers must emphasize operational safety. They may spend years searching satellite images for the safest landing sites that satisfy mission objectives. And they have a lot of ground to cover. Mars has almost the same amount of dry land as Earth!

Analyzing an area this large requires the help of computers. In this chapter, you'll use Python and the Jet Propulsion Laboratory's pride and joy, the Mars Orbiter Laser Altimeter (MOLA) map, to choose and rank candidate landing sites for a Mars lander. To load and extract useful information from the MOLA map, you'll use the Python Imaging Library, OpenCV, tkinter, and `NumPy`.

How to Land on Mars

There are many ways to land a probe on Mars, including with parachutes, balloons, retro rockets, and jet packs. Regardless of the method, most landings follow the same basic safety rules.

The first rule is to target low-lying areas. A probe may enter the Martian atmosphere going as fast as 27,000 kilometers per hour (kph). Slowing it down for a soft landing requires a nice thick atmosphere. But the Martian atmosphere is thin—roughly 1 percent the density of Earth's. To find enough of it to make a difference, you need to aim for the lowest elevations, where the air is denser and the flight through it takes as long as possible.

Unless you have a specialty probe, like one designed for a polar cap, you'll want to land near the equator. Here, you'll find plenty of sunshine to feed the probe's solar panels, and temperatures stay warm enough to protect the probe's delicate machinery.

You'll want to avoid sites covered in boulders that can destroy the probe, prevent its panels from opening, block its robotic arm, or leave it tilted away from the sun. For similar reasons, you'll want to stay away from areas with steep slopes, such as those found on the rims of craters. From a safety standpoint, flatter is better, and boring is beautiful.

Another challenge of landing on Mars is that you can't be very precise. It's hard to fly 50 million kilometers or more, graze the atmosphere, and land exactly where you intended. Inaccuracies in interplanetary navigation, along with variances in Martian atmospheric properties, make hitting a small target very uncertain.

Consequently, NASA runs lots of computer simulations for each landing coordinate. Each simulation run produces a coordinate, and the scatter of points that results from thousands of runs forms an elliptical shape with the long axis parallel to the probe's flight path. These *landing ellipses* can be quite large (Figure 7-1), though the accuracy improves with each new mission.

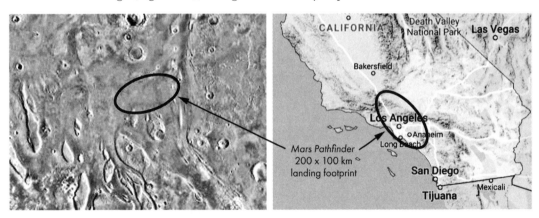

Figure 7-1: Scaled comparison of 1997 Mars Pathfinder landing site (left) with Southern California (right)

The 2018 *InSight* lander had a landing ellipse of only 130 km × 27 km. The probability of the probe landing somewhere within that ellipse was around 99 percent.

The MOLA Map

To identify suitable landing spots, you'll need a map of Mars. Between 1997 and 2001, a tool aboard the *Mars Global Surveyor* (*MGS*) spacecraft shined a laser on Mars and timed its reflection 600 million times. From these measurements, researchers led by Maria Zuber and David Smith produced a detailed global topography map known as MOLA (Figure 7-2).

Figure 7-2: MOLA shaded relief map of Mars

To see the spectacular color version of MOLA, along with a legend, go to the Wikipedia page for the *Mars Global Surveyor*. The blues in this map correspond to where oceans and seas probably existed on Mars billions of years ago. Their distribution is based on a combination of elevation and diagnostic surface features, like ancient shorelines.

The laser measurements for MOLA have a vertical positional accuracy of around 3 to 13 m and a horizontal positional accuracy of about 100 m. Pixel resolution is 463 m per pixel. By itself, the MOLA map lacks the detail needed to safely choose a final landing ellipse, but it's perfect for the scoping work you'll be asked to do.

Project #10: Selecting Martian Landing Sites

Let's pretend you're a NASA summer intern working on the Orpheus Project, a mission designed to listen for marsquakes and study the interior of the planet, much like the 2018 Mars *InSight* mission. Because the purpose of Orpheus is to study the interior of Mars, interesting features of the planet's surface aren't that important. Safety is the prime concern, making this mission an engineer's dream come true.

Your job is to find at least a dozen regions from which NASA staff can select smaller candidate landing ellipses. According to your supervisor, the regions should be rectangles 670 km long (E–W) and 335 km wide (N–S). To address safety concerns, the regions should straddle the equator between 30° N and 30° S latitude, lie at low elevations, and be as smooth and flat as possible.

THE OBJECTIVE

Write a Python program that uses an image of the MOLA map to choose the 20 safest 670 km × 335 km regions near the Martian equator from which to select landing ellipses for the *Orpheus* lander.

The Strategy

First, you'll need a way to divide the MOLA digital map into rectangular regions and extract statistics on elevation and surface roughness. This means you'll be working with pixels, so you'll need imaging tools. And since NASA is always containing costs, you'll want to use free, open source libraries like OpenCV, the Python Imaging Library (PIL), tkinter, and NumPy. For an overview and installation instructions, see "Installing the Python Libraries" on page 6 for OpenCV and NumPy, and see "The Word Cloud and PIL Modules" on page 65 for PIL. The tkinter module comes preinstalled with Python.

To honor the elevation constraints, you can simply calculate the average elevation for each region. For measuring how smooth a surface is at a given scale, you have lots of choices, some of them quite sophisticated. Besides basing smoothness on elevation data, you can look for differential shadowing in stereo images; the amount of scattering in radar, laser, and microwave reflections; thermal variations in infrared images; and so on. Many roughness estimates involve tedious analyses along *transects*, which are lines drawn on the planet's surface along which variations in height are measured and scrutinized. Since you're not really a summer intern with three months to burn, you're going to keep things simple and use two common measurements that you'll apply to each rectangular region: standard deviation and peak-to-valley.

Standard deviation, also called *root-mean-square* by physical scientists, is a measure of the spread in a set of numbers. A low standard deviation indicates that the values in a set are close to the average value; a high standard deviation indicates they are spread out over a wider range. A map region with a low standard deviation for elevation means that the area is flattish, with little variance from the average elevation value.

Technically, the standard deviation for a population of samples is the square root of the average of the squared deviations from the mean, represented by the following formula:

$$\sigma = \sqrt{\frac{1}{N} \sum_{i=1}^{N} \left(h_i - h_0 \right)^2}$$

where σ is the standard deviation, N is the number of samples, h_i is the current height sample, and h_0 is the mean of all the heights.

The *peak-to-valley* statistic is the difference in height between the highest and lowest points on a surface. It captures the maximum elevation change for the surface. This is important as a surface may have a relatively low standard deviation—suggesting smoothness—yet contain a significant hazard, as shown in the cross section in Figure 7-3.

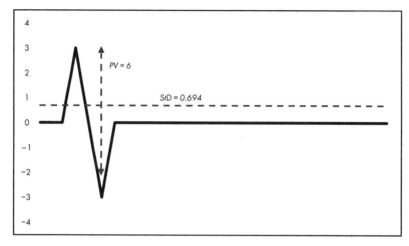

Figure 7-3: A surface profile (black line) with standard deviation (StD) and peak-to-valley (PV) statistics

You can use the standard deviation and peak-to-valley statistics as comparative metrics. For each rectangular region, you're looking for the lowest values of each statistic. And because each statistic records something slightly different, you'll find the best 20 rectangular regions based on each statistic and then select only the rectangles that overlap to find the best rectangles overall.

The Site Selector Code

The *site_selector.py* program uses a grayscale image of the MOLA map (Figure 7-4) to select the landing site rectangles and the shaded color map (Figure 7-2) to post them. Elevation is represented by a single channel in the grayscale image, so it's easier to use than the three-channel (RGB) color image.

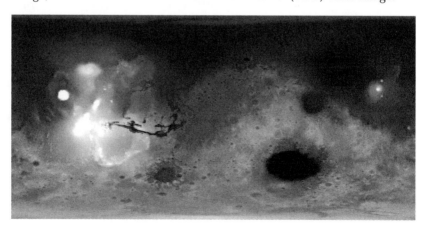

Figure 7-4: Mars MGS MOLA Digital Elevation Model 463m v2 (mola_1024x501.png)

You can find the program, the grayscale image (*mola_1024x501.png*), and the color image (*mola_color_1024x506.png*) in the *Chapter_7* folder, downloadable from *https://nostarch.com/real-world-python/*. Keep these files together in the same folder and don't rename them.

NOTE   *The MOLA map comes in multiple file sizes and resolutions. You're using the smallest size here to speed up the download and run times.*

Importing Modules and Assigning User Input Constants

Listing 7-1 imports modules and assigns constants that represent user input parameters. These include image filenames, the dimensions of the rectangular regions, a maximum elevation limit, and the number of candidate rectangles to consider.

site_selector.py, part 1

```
import tkinter as tk
from PIL import Image, ImageTk
import numpy as np
import cv2 as cv

# CONSTANTS: User Input:
IMG_GRAY = cv.imread('mola_1024x501.png', cv.IMREAD_GRAYSCALE)
IMG_COLOR = cv.imread('mola_color_1024x506.png')
RECT_WIDTH_KM = 670
RECT_HT_KM = 335
MAX_ELEV_LIMIT = 55
NUM_CANDIDATES = 20
MARS_CIRCUM = 21344
```

Listing 7-1: Importing modules and assigning user input constants

Start by importing the `tkinter` module. This is Python's default GUI library for developing desktop applications. You'll use it to make the final display: a window with the color MOLA map at the top and a text description of the posted rectangles at the bottom. Most Windows, macOS, and Linux machines come with `tkinter` already installed. If you don't have it or need the latest version, you can download and install it from *https://www.activestate.com/*. Online documentation for the module can be found at *https://docs.python.org/3/library/tk.html*.

Next, import the `Image` and `ImageTK` modules from the Python Imaging Library. The `Image` module provides a class that represents a PIL image. It also provides factory functions, including functions to load images from files and create new images. The `ImageTK` module contains support for creating and modifying `tkinter`'s `BitmapImage` and `PhotoImage` objects from PIL images. Again, you'll use these at the end of the program to place the color map and some descriptive text in a summary window. Finally, finish the imports with NumPy and OpenCV.

Now, assign some constants that represent user input that won't change as the program runs. First, use the OpenCV `imread()` method to load the grayscale MOLA image. Note that you have to use the `cv.IMREAD_GRAYSCALE` flag, as the method loads images in color by default. Repeat the code without the flag to load the color image. Then add constants for the rectangle size. In the next listing, you'll convert these dimensions to pixels for use with the map image.

Next, to ensure the rectangles target smooth areas at low elevations, you should limit the search to lightly cratered, flat terrain. These regions are believed to represent old ocean bottoms. Thus, you'll want to set the maximum elevation limit to a grayscale value of 55, which corresponds closely to the areas thought to be remnants of ancient shorelines (see Figure 7-5).

Figure 7-5: MOLA map with pixel values ≤ 55 colored black to represent ancient Martian oceans

Now, specify the number of rectangles to display, represented by the NUM_CANDIDATES variable. Later, you'll select these from a sorted list of rectangle statistics. Complete the user input constants by assigning a constant to hold the Martian circumference, in kilometers. You'll use this later to determine the number of pixels per kilometer.

Assigning Derived Constants and Creating the screen Object

Listing 7-2 assigns constants that are derived from other constants. These values will update automatically if the user changes the previous constants, for example, to test different rectangle sizes or elevation limits. The listing ends by creating tkinter screen and canvas objects for the final display.

site_selector.py, part 2

```
# CONSTANTS: Derived:
IMG_HT, IMG_WIDTH = IMG_GRAY.shape
PIXELS_PER_KM = IMG_WIDTH / MARS_CIRCUM
RECT_WIDTH = int(PIXELS_PER_KM * RECT_WIDTH_KM)
RECT_HT = int(PIXELS_PER_KM * RECT_HT_KM)
❶ LAT_30_N = int(IMG_HT / 3)
LAT_30_S = LAT_30_N * 2
STEP_X = int(RECT_WIDTH / 2)
STEP_Y = int(RECT_HT / 2)

❷ screen = tk.Tk()
canvas = tk.Canvas(screen, width=IMG_WIDTH, height=IMG_HT + 130)
```

Listing 7-2: Assigning derived constants and setting up the tkinter screen

Start by unpacking the height and width of the image using the shape attribute. OpenCV stores images as NumPy ndarrays, which are *n*-dimensional arrays—or tables—of elements of the same type. For an image array, shape is a tuple of the number of rows, columns, and channels. The height represents the number of pixel rows in the image, and the width represents the number of pixel columns in the image. Channels represent the number of components used to represent each pixel (such as red, green, and blue). For grayscale images with one channel, shape is just a tuple of the area's height and width.

To convert the rectangle dimensions from kilometers to pixels, you need to know how many pixels there are per kilometer. So, divide the image width by the circumference to get the pixels per kilometer at the equator. Then convert width and height into pixels. You'll use these to derive values for index slicing later, so make sure they are integers by using int(). The value of these constants should now be 32 and 16, respectively.

You want to limit your search to the warmest and sunniest areas, which straddle the equator between 30° north latitude and 30° south latitude (Figure 7-6). In terms of climatic criteria, this region corresponds to the tropics on Earth.

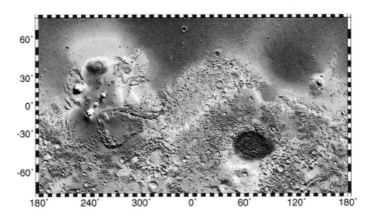

Figure 7-6: Latitude (y-axis) and longitude (x-axis) on Mars

Latitude values start at 0° at the equator and end at 90° at the poles. To find 30° north, all you need to do is divide the image height by 3 ❶. To get to 30° south, double the number of pixels it took to get to 30° north.

Restricting the search to the equatorial region of Mars has a beneficial side effect. The MOLA map you're using is based on a *cylindrical projection*, used to transfer the surface of a globe onto a flat plane. This causes converging lines of longitude to be parallel, badly distorting features near the poles. You may have noticed this on wall maps of the earth, where Greenland looks like a continent and Antarctica is impossibly huge (see Figure 7-7).

Fortunately, this distortion is minimized near the equator, so you don't have to factor it into the rectangle dimensions. You can verify this by checking the shape of craters on the MOLA map. So long as they're nice and circular—rather than oval—projection-related effects can be ignored.

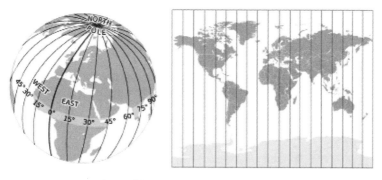

Figure 7-7: Forcing lines of longitude to be parallel distorts the size of features near the poles.

Next, you'll need to divide up the map into rectangular regions. A logical place to begin is the upper-left corner, tucked under the 30° north latitude line (Figure 7-8).

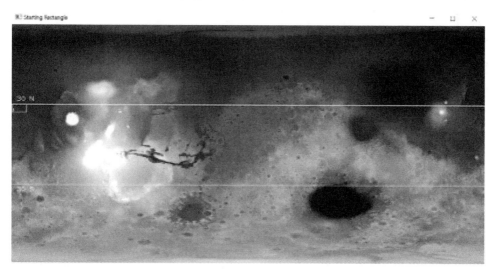

Figure 7-8: Position of the first numbered rectangle

The program will draw this first rectangle, number it, and calculate the elevation statistics within it. It will then move the rectangle eastward and repeat the process. How far you move the rectangle each time is defined by the STEP_X and STEP_Y constants and depends on something called *aliasing*.

Aliasing is a resolution issue. It occurs when you don't take enough samples to identify all the important surface features in an area. This can cause you to "skip over" a feature, such as a crater, and fail to recognize it. For example, in Figure 7-9A, there's a suitably smooth landing ellipse between two large craters. However, as laid out in Figure 7-9B, no rectangular region corresponds to this ellipse; both rectangles in the vicinity partially sample a crater rim. As a result, none of the drawn rectangles contains a suitable landing ellipse, even though one exists in the vicinity.

With this arrangement of rectangles, the ellipse in Figure 7-9A is *aliased*. But shift each rectangle by half its width, as in Figure 7-9C, and the smooth area is properly sampled and recognized.

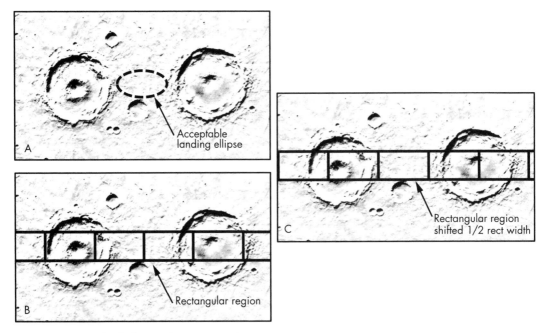

Figure 7-9: Example of aliasing due to rectangle positioning

The rule of thumb to avoid aliasing effects is to make the step size less than or equal to half the width of the smallest feature you want to identify. For this project, use half the rectangle width so the displays don't become too busy.

Now it's time to look ahead to the final display. Create a screen instance of the tkinter Tk() class ❷. The tkinter application is Python's wrapper of the GUI toolkit Tk, originally written in a computer language called TCL. It needs the screen window to link to an underlying tcl/tk interpreter that translates tkinter commands into tcl/tk commands.

Next, create a tkinter canvas object. This is a rectangular drawing area designed for complex layouts of graphics, text, widgets, and frames. Pass it the screen object, set its width equal to the MOLA image, and set its height equal to the height of the MOLA image plus 130. The extra padding beneath the image will hold the text summarizing the statistics for the displayed rectangles.

It's more typical to place the tkinter code just described at the *end* of programs, rather than at the beginning. I chose to put it near the top to make the code explanation easier to follow. You can also embed this code within the function that makes the final display. However, this can cause problems for macOS users. For macOS 10.6 or newer, the Apple-supplied Tcl/Tk 8.5 has serious bugs that can cause application crashes (see *https:// www.python.org/download/mac/tcltk/*).

Defining and Initializing a Search Class

Listing 7-3 defines a class that you'll use to search for suitable rectangular regions. It then defines the class's __init__() initialization method, used to instantiate new objects. For a quick overview of OOP, see "Defining the Search Class" on page 10, where you also define a search class.

site_selector.py, part 3

```
class Search():
    """Read image and identify landing rectangles based on input criteria."""

    def __init__(self, name):
        self.name = name
    ❶ self.rect_coords = {}
        self.rect_means = {}
        self.rect_ptps = {}
        self.rect_stds = {}
    ❷ self.ptp_filtered = []
        self.std_filtered = []
        self.high_graded_rects = []
```

Listing 7-3: Defining the Search class and __init__() method

Define a class called Search. Then define the __init__() method used to create new objects. The name parameter will allow you to give a personalized name to each object when you create it later in the main() function.

Now you're ready to start assigning attributes. Start by linking the object's name with the argument you'll provide when you create the object. Then assign four empty dictionaries to hold important statistics for each rectangle ❶. These include the rectangle's corner-point coordinates and its mean elevation, peak-to-valley, and standard deviation statistics. For a key, all these dictionaries will use consecutive numbers, starting with 1. You'll want to filter the statistics to find the lowest values, so set up two empty lists to hold these ❷. Note that I use the term ptp, rather than ptv, to represent the peak-to-valley statistic. That's to be consistent with the NumPy built-in method for this calculation, which is called *peak-to-peak*.

At the end of the program, you'll place rectangles that occur in both the sorted standard deviation and peak-to-valley lists in a new list named high_graded_rects. This list will contain the numbers of the rectangles with the lowest combined scores. These rectangles will be the best places to look for landing ellipses.

Calculating Rectangle Statistics

Still in the Search class, Listing 7-4 defines a method that calculates statistics in a rectangle, adds the statistics to the appropriate dictionary, and then moves to the next rectangle and repeats the process. The method honors the elevation limit by using the rectangles in low-lying areas only to populate the dictionaries.

```
def run_rect_stats(self):
    """Define rectangular search areas and calculate internal stats."""
    ul_x, ul_y = 0, LAT_30_N
    lr_x, lr_y = RECT_WIDTH, LAT_30_N + RECT_HT
    rect_num = 1

    while True:
 ❶      rect_img = IMG_GRAY[ul_y : lr_y, ul_x : lr_x]
        self.rect_coords[rect_num] = [ul_x, ul_y, lr_x, lr_y]
        if np.mean(rect_img) <= MAX_ELEV_LIMIT:
            self.rect_means[rect_num] = np.mean(rect_img)
            self.rect_ptps[rect_num] = np.ptp(rect_img)
            self.rect_stds[rect_num] = np.std(rect_img)
        rect_num += 1

        ul_x += STEP_X
        lr_x = ul_x + RECT_WIDTH
 ❷      if lr_x > IMG_WIDTH:
            ul_x = 0
            ul_y += STEP_Y
            lr_x = RECT_WIDTH
            lr_y += STEP_Y
 ❸      if lr_y > LAT_30_S + STEP_Y:
            break
```

Listing 7-4: Calculating rectangle statistics and moving the rectangle

Define the run_rect_stats() method, which takes self as an argument. Then assign local variables for the upper-left and lower-right corners of each rectangle. Initialize them using a combination of coordinates and constants. This will place the first rectangle along the left side of the image with its top boundary at 30° north latitude.

Keep track of the rectangles by numbering them, starting with 1. These numbers will serve as the keys for the dictionaries used to record coordinates and stats. You'll also use them to identify the rectangles on the map, as demonstrated earlier in Figure 7-8.

Now, start a while loop that will automate the process of moving the rectangles and recording their statistics. This loop will run until more than half of a rectangle extends below latitude 30° south, at which time the loop will break.

As mentioned previously, OpenCV stores images as NumPy arrays. To calculate the stats within the active rectangle, rather than the whole image, create a subarray using normal slicing ❶. Call this subarray rect_img, for "rectangular image." Then, add the rectangle number and these coordinates to the rect_coords dictionary. You'll want to keep a record of these coordinates for the NASA staff, who'll use your rectangles as the starting point for more detailed investigations later.

Next, start a conditional to check that the current rectangle is at or below the maximum elevation limit specified for the project. As part of this statement, use NumPy to calculate the mean elevation for the rect_img subarray.

If the rectangle passes the elevation test, populate the three dictionaries with the coordinates, peak-to-valley, and standard deviation statistics, as appropriate. Note that you can perform the calculation as part of the process, using np.ptp for peak-to-valley and np.std for standard deviation.

Next, advance the rect_num variable by 1 and move the rectangle. Move the upper-left x-coordinate by the step size and then shift the lower-right x-coordinate by the width of the rectangle. You don't want the rectangle to extend past the right side of the image, so check whether lr_x is greater than the image width ❷. If it is, set the upper-left x-coordinate to 0 to move the rectangle back to the starting position on the left side of the screen. Then move its y-coordinates down so that the new rectangles move along a new row. If the bottom of this new row is more than half a rectangle height below 30° south latitude, you've fully sampled the search area and can end the loop ❸.

Between 30° north and south latitude, the image is bounded on both sides by relatively high, cratered terrain that isn't suitable for a landing site (see Figure 7-6). Thus, you can ignore the final step that shifts the rectangle by one-half its width. Otherwise, you would need to add code that wraps a rectangle from one side of the image to the other and calculates the statistics for each part. We'll take a closer look at this situation in the final challenge project at the end of the chapter.

NOTE   *When you draw something on an image, such as a rectangle, the drawing becomes part of the image. The altered pixels will be included in any NumPy analyses you run, so be sure to calculate any statistics before you annotate the image.*

Checking the Rectangle Locations

Still indented under the Search class, Listing 7-5 defines a method that performs quality control. It prints the coordinates of all the rectangles and then draws them on the MOLA map. This will let you verify that the search area has been fully evaluated and the rectangle size is what you expected it to be.

site_selector.py, part 5

```
def draw_qc_rects(self):
    """Draw overlapping search rectangles on image as a check."""
    img_copy = IMG_GRAY.copy()
    rects_sorted = sorted(self.rect_coords.items(), key=lambda x: x[0])
    print("\nRect Number and Corner Coordinates (ul_x, ul_y, lr_x, lr_y):")
    for k, v in rects_sorted:
        print("rect: {}, coords: {}".format(k, v))
        cv.rectangle(img_copy,
                     (self.rect_coords[k][0], self.rect_coords[k][1]),
                     (self.rect_coords[k][2], self.rect_coords[k][3]),
                     (255, 0, 0), 1)
    cv.imshow('QC Rects {}'.format(self.name), img_copy)
    cv.waitKey(3000)
    cv.destroyAllWindows()
```

Listing 7-5: Drawing all the rectangles on the MOLA map as a quality control step

Start by defining a method to draw the rectangles on the image. Anything you draw on an image in OpenCV becomes part of the image, so first make a copy of the image in the local space.

You'll want to provide NASA with a printout of the identification number and coordinates for each rectangle. To print these in numerical order, sort the items in the rect_coords dictionary using a lambda function. If you haven't used a lambda function before, you can find a short description on page 106 in Chapter 5.

Print a header for the list and then start a for loop through the keys and values in the newly sorted dictionary. The key is the rectangle number, and the value is the list of coordinates, as shown in the following output:

```
Rect Number and Corner Coordinates (ul_x, ul_y, lr_x, lr_y):
rect: 1, coords: [0, 167, 32, 183]
rect: 2, coords: [16, 167, 48, 183]

--snip--

rect: 1259, coords: [976, 319, 1008, 335]
rect: 1260, coords: [992, 319, 1024, 335]
```

Use the OpenCV rectangle() method to draw the rectangles on the image. Pass it the image on which to draw, the rectangle coordinates, a color, and a line width. Access the coordinates directly from the rect_coords dictionary using the key and the list index (0 = upper-left *x*, 1 = upper-left *y*, 2 = lower-right *x*, 3 = lower-right *y*).

To display the image, call the OpenCV imshow() method and pass it a name for the window and the image variable. The rectangles should cover Mars in a band centered on the equator (Figure 7-10). Leave the window up for three seconds and then destroy it.

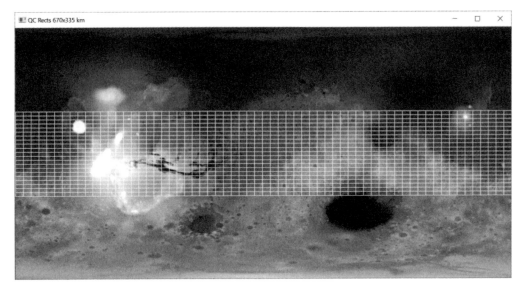

Figure 7-10: All 1,260 rectangles drawn by the draw_qc_rects() method

If you compare Figure 7-10 to Figure 7-8, you may notice that the rectangles appear smaller than expected. This is because you stepped the rectangles across and down the image using half the rectangle width and height so that they overlap each other.

Sorting the Statistics and High Grading the Rectangles

Continuing with the Search class definition, Listing 7-6 defines a method to find the rectangles with the best potential landing sites. The method sorts the dictionaries containing the rectangle statistics, makes lists of the top rectangles based on the peak-to-valley and standard deviation statistics, and then makes a list of any rectangles shared between these two lists. The shared rectangles will be the best candidates for landing sites, as they'll have the smallest peak-to-valley and standard deviation statistics.

site_selector.py, part 6

```
def sort_stats(self):
    """Sort dictionaries by values and create lists of top N keys."""
    ptp_sorted = (sorted(self.rect_ptps.items(), key=lambda x: x[1]))
    self.ptp_filtered = [x[0] for x in ptp_sorted[:NUM_CANDIDATES]]
    std_sorted = (sorted(self.rect_stds.items(), key=lambda x: x[1]))
    self.std_filtered = [x[0] for x in std_sorted[:NUM_CANDIDATES]]
    for rect in self.std_filtered:
        if rect in self.ptp_filtered:
            self.high_graded_rects.append(rect)
```

Listing 7-6: Sorting and high grading the rectangles based on their statistics

Define a method called sort_stats(). Sort the rect_ptps dictionary with a lambda function that sorts the values rather than the keys. The values in this dictionary are the peak-to-valley measurements. This should create a list of tuples, with the rectangle number at index 0 and the peak-to-valley value at index 1.

Next, use list comprehension to populate the self.ptp_filtered attribute with the rectangle numbers in the ptp_sorted list. Use index slicing to select only the first 20 values, as stipulated by the NUM_CANDIDATES constant. You now have the 20 rectangles with the lowest peak-to-valley scores. Repeat this same basic code for standard deviation, producing a list of the 20 rectangles with the lowest standard deviation.

Finish the method by looping through the rectangle numbers in the std_filtered list and comparing them to those in the ptp_filtered list. Append matching numbers to the high_graded_rects instance attribute you created previously with the __init__() method.

Drawing the Filtered Rectangles on the Map

Listing 7-7, still indented under the Search class, defines a method that draws the 20 best rectangles on the grayscale MOLA map. You'll call this method in the main() function.

```
def draw_filtered_rects(self, image, filtered_rect_list):
    """Draw rectangles in list on image and return image."""
    img_copy = image.copy()
    for k in filtered_rect_list:
        cv.rectangle(img_copy,
                     (self.rect_coords[k][0], self.rect_coords[k][1]),
                     (self.rect_coords[k][2], self.rect_coords[k][3]),
                     (255, 0, 0), 1)
        cv.putText(img_copy, str(k),
                   (self.rect_coords[k][0] + 1, self.rect_coords[k][3]- 1),
                   cv.FONT_HERSHEY_PLAIN, 0.65, (255, 0, 0), 1)

❶   cv.putText(img_copy, '30 N', (10, LAT_30_N - 7),
               cv.FONT_HERSHEY_PLAIN, 1, 255)
    cv.line(img_copy, (0, LAT_30_N), (IMG_WIDTH, LAT_30_N),
            (255, 0, 0), 1)
    cv.line(img_copy, (0, LAT_30_S), (IMG_WIDTH, LAT_30_S),
            (255, 0, 0), 1)
    cv.putText(img_copy, '30 S', (10, LAT_30_S + 16),
               cv.FONT_HERSHEY_PLAIN, 1, 255)

    return img_copy
```

Listing 7-7: Drawing filtered rectangles and latitude lines on MOLA map

Start by defining the method, which in this case takes multiple arguments. Besides self, the method will need a loaded image and a list of rectangle numbers. Use a local variable to copy the image and then start looping through the rectangle numbers in the filtered_rect_list. With each loop, draw a rectangle by using the rectangle number to access the corner coordinates in the rect_coords dictionary.

So you can tell one rectangle from another, use OpenCV's putText() method to post the rectangle number in the bottom-left corner of each rectangle. It needs the image, the text (as a string), coordinates for the upper-left *x* and lower-right *x*, a font, a line width, and a color.

Next, draw the annotated latitude limits, starting with the text for 30° north ❶. Then draw the line using OpenCV's line() method. It takes as arguments an image, a pair of (*x*, *y*) coordinates for the start and end of the line, a color, and a thickness. Repeat these basic instructions for 30° south latitude.

End the method by returning the annotated image. The best rectangles, based on the peak-to-valley and standard deviation statistics, are shown in Figures 7-11 and 7-12, respectively.

These two figures show the top 20 rectangles for each statistic. That doesn't mean they always agree. The rectangle with the lowest standard deviation may not appear in the peak-to-valley figure due to the presence of a single small crater. To find the flattest, smoothest rectangles, you need to identify the rectangles that appear in both figures and show them in their own display.

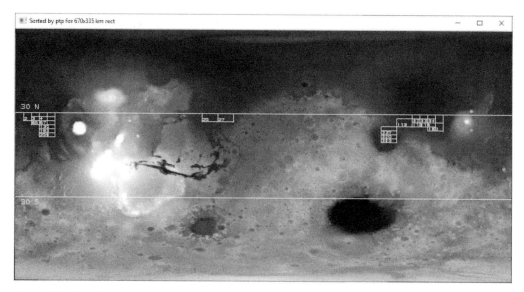

Figure 7-11: The 20 rectangles with the lowest peak-to-valley scores

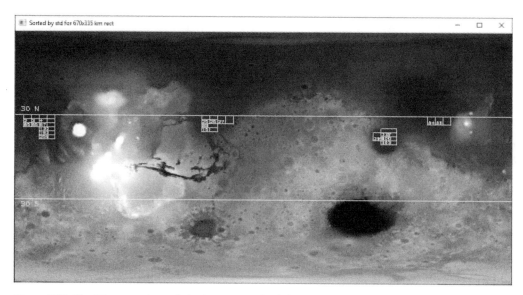

Figure 7-12: The 20 rectangles with the lowest standard deviations

Making the Final Color Display

Listing 7-8 finishes the Search class by defining a method to summarize the best rectangles. It uses tkinter to make a summary window with the rectangles posted on the color MOLA image. It also prints the rectangles' statistics below the image as text objects. This adds a little work, but it's a cleaner-looking solution than posting the summarized stats directly on the image with OpenCV.

```
def make_final_display(self):
    """Use Tk to show map of final rects & printout of their statistics."""
    screen.title('Sites by MOLA Gray STD & PTP {} Rect'.format(self.name))

    img_color_rects = self.draw_filtered_rects(IMG_COLOR,
                                               self.high_graded_rects)

❶  img_converted = cv.cvtColor(img_color_rects, cv.COLOR_BGR2RGB)
    img_converted = ImageTk.PhotoImage(Image.fromarray(img_converted))
    canvas.create_image(0, 0, image=img_converted, anchor=tk.NW)

❷  txt_x = 5
    txt_y = IMG_HT + 20
    for k in self.high_graded_rects:
        canvas.create_text(txt_x, txt_y, anchor='w', font=None,
                          text="rect={} mean elev={:.1f} std={:.2f} ptp={}"
                          .format(k, self.rect_means[k], self.rect_stds[k],
                                  self.rect_ptps[k]))
        txt_y += 15
      ❸ if txt_y >= int(canvas.cget('height')) - 10:
            txt_x += 300
            txt_y = IMG_HT + 20
    canvas.pack()
    screen.mainloop()
```

Listing 7-8: Making the final display using the color MOLA map

After defining the method, give the tkinter screen window a title that links to the name of your search object.

Then, to make the final color image for display, name a local variable img_color_rects and call the draw_filtered_rects() method. Pass it the color MOLA image and the list of high-graded rectangles. This will return the colored image with the final rectangles and latitude limits.

Before you can post this new color image in the tkinter canvas, you need to convert the colors from OpenCV's Blue-Green-Red (BGR) format to the Red-Green-Blue (RGB) format used by tkinter. Do this with the OpenCV cvtColor() method. Pass it the image variable and the COLOR_BGR2RGB flag ❶. Name the result img_converted.

At this point, the image is still a NumPy array. To convert to a tkinter-compatible photo image, you need to use the PIL ImageTk module's PhotoImage class and the Image module's fromarray() method. Pass the method the RGB image variable you created in the previous step.

With the image finally tkinter ready, place it in the canvas using the create_image() method. Pass the method the coordinates of the upper-left corner of the canvas (0, 0), the converted image, and a northwest anchor direction.

Now all that's left is to add the summary text. Start by assigning coordinates for the bottom-left corner of the first text object ❷. Then begin looping through the rectangle numbers in the high-graded rectangle list. Use the create_text() method to place the text in the canvas. Pass it a pair of coordinates, a left-justified anchor direction, the default font, and a text string. Get the statistics by accessing the different dictionaries using the rectangle number, designated k for "key."

Increment the text box's y-coordinate by 15 after drawing each text object. Then write a conditional to check that the text is greater than or within 10 pixels of the bottom of the canvas ❸. You can obtain the height of the canvas using the cget() method.

If the text is too close to the bottom of the canvas, you need to start a new column. Shift the txt_x variable over by 300 and reset txt_y to the image height plus 20.

Finish the method definition by packing the canvas and then calling the screen object's mainloop(). Packing optimizes the placement of objects in the canvas. The mainloop() is an infinite loop that runs tkinter, waits for an event to occur, and processes the event until the window is closed.

NOTE   *The height of the color image (506 pixels) is slightly larger than that of the grayscale image (501 pixels). I chose to ignore this, but if you're a stickler for accuracy, you can use OpenCV to shrink the height of the color image using IMG_COLOR = cv.resize (IMG_COLOR, (1024, 501), interpolation = cv.INTER_AREA).*

Running the Program with main()

Listing 7-9 defines a main() function to run the program.

site_selector.py, part 9

```
def main():
    app = Search('670x335 km')
    app.run_rect_stats()
    app.draw_qc_rects()
    app.sort_stats()
    ptp_img = app.draw_filtered_rects(IMG_GRAY, app.ptp_filtered)
    std_img = app.draw_filtered_rects(IMG_GRAY, app.std_filtered)

❶  cv.imshow('Sorted by ptp for {} rect'.format(app.name), ptp_img)
    cv.waitKey(3000)
    cv.imshow('Sorted by std for {} rect'.format(app.name), std_img)
    cv.waitKey(3000)

    app.make_final_display()  # Includes call to mainloop().

❷ if __name__ == '__main__':
    main()
```

Listing 7-9: Defining and calling the main() function used to run the program

Start by instantiating an app object from the Search class. Name it 670x335 km to document the size of the rectangular regions being investigated. Next, call the Search methods in order. Run the statistics on the rectangles and draw the quality control rectangles. Sort the statistics from smallest to largest and then draw the rectangles with the best peak-to-valley and standard deviation statistics. Show the results ❶ and finish the function by making the final summary display.

Back in the global space, add the code that lets the program run as an imported module or in stand-alone mode ❷.

Figure 7-13 shows the final display. It includes the high-graded rectangles and the summary statistics sorted based on standard deviation.

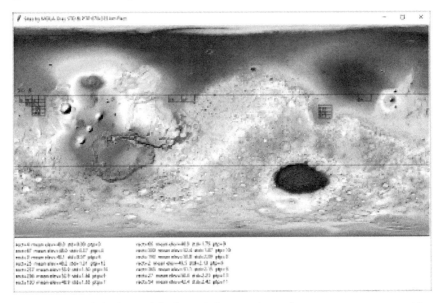

Figure 7-13: Final display with high-graded rectangles and summary statistics sorted by standard deviation

Results

After you've made the final display, the first thing you should do is perform a sanity check. Make sure that the rectangles are within the allowed latitude and elevation limits and that they appear to be in smooth terrain. Likewise, the rectangles based on the peak-to-valley and standard deviation statistics, shown in Figures 7-11 and 7-12, respectively, should match the constraints and mostly pick the same rectangles.

As noted previously, the rectangles in Figures 7-11 and 7-12 don't perfectly overlap. That's because you're using two different metrics for smoothness. One thing you can be sure of, though, is that the rectangles that do overlap will be the smoothest of all the rectangles.

While all the rectangle locations look reasonable in the final display, the concentration of rectangles on the far-west side of the map is particularly encouraging. This is the smoothest terrain in the search area (Figure 7-14), and your program clearly recognized it.

This project focused on safety concerns, but scientific objectives drive site selection for most missions. In the practice projects at the end of the chapter, you'll get a chance to incorporate an additional constraint—geology—into the site selection equation.

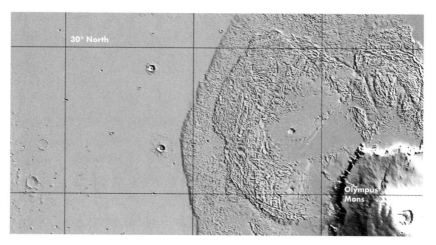

Figure 7-14: The very smooth terrain west of the Olympus Mons lava fields

Summary

In this chapter, you used Python, OpenCV, the Python Imaging Library, NumPy, and tkinter to load, analyze, and display an image. Because OpenCV treats images as NumPy arrays, you can easily extract information from parts of an image and evaluate it with Python's many scientific libraries.

The dataset you used was quick to download and fast to run. While a real intern would have used a larger and more rigorous dataset, such as one composed of millions of actual elevation measurements, you got to see how the process works with little effort and reasonable results.

Further Reading

The Jet Propulsion Laboratory has several short and fun videos about landing on Mars. Find them with online searches for *Mars in a Minute: How Do You Choose a Landing Site?*, *Mars in a Minute: How Do You Get to Mars?*, and *Mars in a Minute: How Do You Land on Mars?*.

Mapping Mars: Science, Imagination, and the Birth of a World (Picador, 2002), by Oliver Morton, tells the story of the contemporary exploration of Mars, including the creation of the MOLA map.

The Atlas of Mars: Mapping Its Geography and Geology (Cambridge University Press, 2019), by Kenneth Coles, Kenneth Tanaka, and Philip Christensen, is a spectacular all-purpose reference atlas of Mars that includes maps of topography, geology, mineralogy, thermal properties, near-surface water-ice, and more.

The data page for the MOLA map used in Project 10 is at *https://astrogeology.usgs.gov/search/map/Mars/GlobalSurveyor/MOLA/Mars_MGS _MOLA_DEM_mosaic_global_463m/*.

Detailed Martian datasets are available on the Mars Orbital Data Explorer site produced by the PDS Geoscience Node at Washington University in St. Louis (*https://ode.rsl.wustl.edu/mars/index.aspx*).

Practice Project: Confirming That Drawings Become Part of an Image

Write a Python program that verifies that drawings added to an image, such as text, lines, rectangles, and so on, become part of that image. Use NumPy to calculate the mean, standard deviation, and peak-to-valley statistics on a rectangular region in the MOLA grayscale image, but don't draw the rectangle outline. Then draw a white line around the region and rerun the statistics. Do the two runs agree?

You can find a solution, *practice_confirm_drawing_part_of_image.py*, in the appendix or *Chapter_7* folder, downloadable from *https://nostarch.com/real-world-python/*.

Practice Project: Extracting an Elevation Profile

An elevation profile is a two-dimensional, cross-sectional view of a landscape. It provides a side view of a terrain's relief along a line drawn between locations on a map. Geologists can use profiles to study the smoothness of a surface and visualize its topography. For this practice project, draw a west-to-east profile that passes through the caldera of the largest volcano in the solar system, Olympus Mons (Figure 7-15).

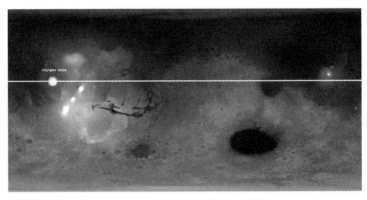

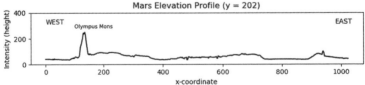

Figure 7-15: Vertically exaggerated west-east profile through Olympus Mons

Use the *Mars MGS MOLA - MEX HRSC Blended DEM Global 200m v2* map shown in Figure 7-15. This version has better lateral resolution than the one you used for Project 10. It also uses the full elevation range in the MOLA data. You can find a copy, *mola_1024x512_200mp.jpg*, in the *Chapter_7* folder, downloadable from the book's website. A solution, *practice_profile_olympus.py*, is available in the same folder and in the appendix.

Practice Project: Plotting in 3D

Mars is an asymmetrical planet, with the southern hemisphere dominated by ancient cratered highlands and the north characterized by smooth, flat lowlands. To make this more apparent, use the 3D plotting functionality in `matplotlib` to display the *mola_1024x512_200mp.jpg* image you used in the previous practice project (Figure 7-16).

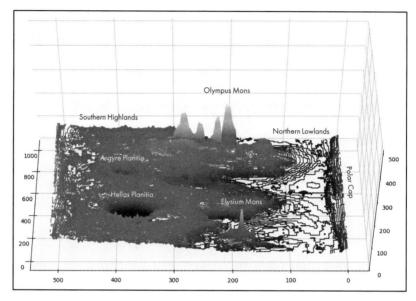

Figure 7-16: A 3D contour plot of Mars, looking toward the west

With `matplotlib`, you can make 3D relief plots using points, lines, contours, wireframes, and surfaces. Although the plots are somewhat crude, you can generate them quickly. You can also use the mouse to interactively grab the plot and change the viewpoint. They are particularly useful for people who have trouble visualizing topography from 2D maps.

In Figure 7-16, the exaggerated vertical scale makes the elevation difference from south to north easy to see. It's also easy to spot the tallest mountain (Olympus Mons) and the deepest crater (Hellas Planitia).

You can reproduce the plot in Figure 7-16—sans annotation—with the *practice_3d_plotting.py* program in the appendix or *Chapter_7* folder, downloadable from the book's website. The map image can be found in the same folder.

Practice Project: Mixing Maps

Make up a new project that adds a bit of science to the site selection process. Combine the MOLA map with a color geology map and find the smoothest rectangular regions within the volcanic deposits at Tharsis Montes (see arrow in Figure 7-17).

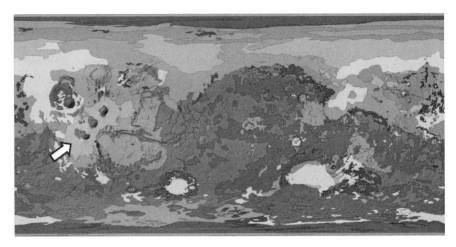

Figure 7-17: Geological map of Mars. The arrow points to the Tharsis volcanic deposits.

Since the Tharsis Montes region lies at a high altitude, focus on finding the flattest and smoothest parts of the volcanic deposits, rather than targeting the lowest elevations. To isolate the volcanic deposits, consider thresholding a grayscale version of the map. *Thresholding* is a segmentation technique that partitions an image into a foreground and a background.

With thresholding, you convert a grayscale image into a binary image where pixels above or between specified threshold values are set to 1 and all others are set to 0. You can use this binary image to filter the MOLA map, as shown in Figure 7-18.

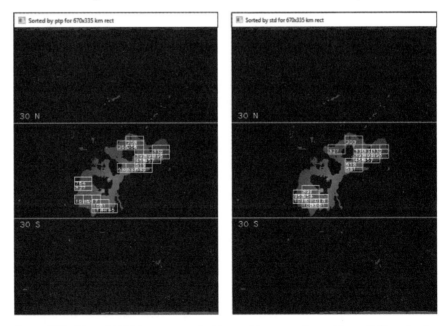

Figure 7-18: Filtered MOLA map over the Tharsis Montes region, with ptp (left) and std (right) rectangles

You can find the geological map, *Mars_Global_Geology_Mariner9_1024.jpg*, in the *Chapter_7* folder, downloadable from the book's website. The volcanic deposits will be light pink in color. For the elevation map, use *mola_1024x512_200mp.jpg* from the "Extracting an Elevation Profile" practice project on page 172.

A solution, contained in *practice_geo_map_step_1of2.py* and *practice_geo_map_step_2of2.py*, can be found in the same folder and in the appendix. Run the *practice_geo_map_step_1of2.py* program first to generate the filter for step 2.

Challenge Project: Making It Three in a Row

Edit the "Extracting an Elevation Profile" project so that the profile passes through the three volcanoes on Tharsis Montes, as shown in Figure 7-19.

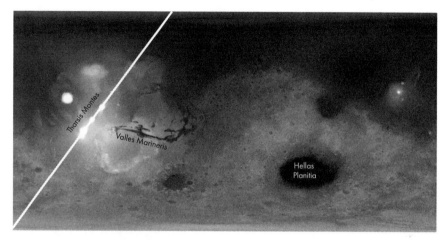

Figure 7-19: Diagonal profile through the three volcanoes on Tharsis Montes

Other interesting features to profile are Valles Marineris, a canyon nine times as long and four times as deep as the Grand Canyon, and Hellas Planitia, considered the third or fourth largest impact crater in the solar system (Figure 7-19).

Challenge Project: Wrapping Rectangles

Edit the *site_selector.py* code so that it accommodates rectangle dimensions that don't divide evenly into the width of the MOLA image. One way to do this is to add code that splits the rectangle into two pieces (one along the right edge of the map and the other along the left edge), calculates statistics for each, and then recombines them into a full rectangle. Another approach is to duplicate the image and "stitch" it to the original image, as shown in Figure 7-20. This way, you won't have to split the rectangles; just decide when to stop stepping them across the map.

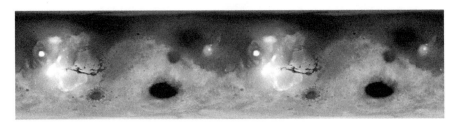

Figure 7-20: The grayscale MOLA image duplicated and repeated

Of course, for efficiency, you don't have to duplicate the whole map. You only need a strip along the eastern margin wide enough to accommodate the final overlapping rectangle.

8

DETECTING DISTANT EXOPLANETS

Extrasolar planets, called exoplanets for short, are planets that orbit alien suns. By the end of 2019, more than 4,000 exoplanets had been discovered. That's an average of 150 per year since the first confirmed discovery in 1992! These days, finding a faraway planet seems as easy as catching a cold, yet it took almost all human history—up to 1930—to discover the eight planets, plus Pluto, that make up our own solar system.

Astronomers detected the first exoplanets by observing gravitationally induced wobble in the motion of stars. Today, they rely mainly on the slight dimming of a star's light as the exoplanet passes between the star and Earth. And with powerful next-generation devices like the James Webb Space Telescope, they'll directly image exoplanets and learn about their rotation, seasons, weather, vegetation, and more.

In this chapter, you'll use OpenCV and `matplotlib` to simulate an exoplanet passing before its sun. You'll record the resulting light curve and then use it to detect the planet and estimate its diameter. Then, you'll simulate how an exoplanet might look to the James Webb Space Telescope. In the "Practice Project" sections, you'll investigate unusual light curves that may represent enormous alien megastructures designed to harness a star's energy.

Transit Photometry

In astronomy, a *transit* occurs when a relatively small celestial body passes directly between the disc of a larger body and an observer. When the small body moves across the face of the larger body, the larger body dims slightly. The best-known transits are those of Mercury and Venus against our own sun (Figure 8-1).

Figure 8-1: Clouds and Venus (the black dot) passing before the sun in June 2012

With today's technology, astronomers can detect the subtle dimming of a faraway star's light during a transit event. The technique, called *transit photometry*, outputs a plot of a star's brightness over time (Figure 8-2).

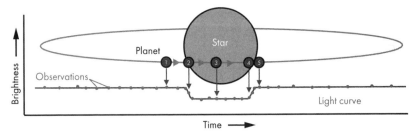

Figure 8-2: The transit photometry technique for detecting exoplanets

In Figure 8-2, the dots on the light curve graph represent measurements of the light given off by a star. When a planet is not positioned over the star ❶, the measured brightness is at a maximum. (We'll ignore light reflected off the exoplanet as it goes through its phases, which would very slightly increase the apparent brightness of the star). As the leading edge of a planet moves onto the disc ❷, the emitted light progressively dims, forming a ramp in the light curve. When the entire planet is visible against the disc ❸, the light curve flattens, and it remains flat until the planet begins exiting the far side of the disc. This creates another ramp ❹, which rises until the planet passes completely off the disc ❺. At that point, the light curve flattens at its maximum value, as the star is no longer obscured.

Because the amount of light blocked during a transit is proportional to the size of the planet's disc, you can calculate the radius of the planet using the following formula:

$$R_p = R_s \sqrt{Depth}$$

where R_p is the planet's radius and R_s is the star's radius. Astronomers determine the star's radius using its distance, brightness, and color, which relates to its temperature. *Depth* refers to the total change in brightness during the transit (Figure 8-3).

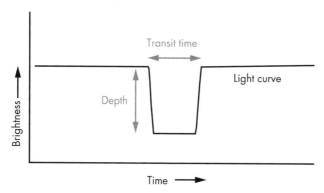

Figure 8-3: Depth represents the total change in brightness observed in a light curve.

Of course, these calculations assume that the whole exoplanet, not just part of it, moved over the face of the star. The latter may occur if the exoplanet skims either the top or bottom of the star (from our point of view). We'll look at this case in "Experimenting with Transit Photometry" on page 182.

Project #11: Simulating an Exoplanet Transit

Before I flew to Idaho to photograph the Great American Eclipse of 2017, I did my homework. The totality event, when the moon completely covered the sun, lasted only 2 minutes and 10 seconds. That left no time for experimenting, testing, or figuring things out on the fly. To successfully capture images of the penumbra, umbra, solar flares, and diamond ring effect (Figure 8-4), I had to know exactly what equipment to take, what camera settings to use, and when these events would occur.

Figure 8-4: Diamond ring effect at the end of totality, 2017 solar eclipse

In a similar fashion, computer simulations prepare you for making observations of the natural world. They help you understand what to expect, when to expect it, and how to calibrate your instruments. In this project, you'll create a simulation of an exoplanet transit event. You can run this simulation with different planet sizes to understand the impact of a transit's size on the light curve. Later, you'll use this simulation to evaluate light curves related to asteroid fields and possible alien megastructures.

THE OBJECTIVE

Write a Python program that simulates an exoplanet transit, plots the resulting light curve, and calculates the radius of the exoplanet.

The Strategy

To generate a light curve, you need to be able to measure changes in brightness. You can do this by performing mathematical operations on pixels, such as finding mean, minimum, and maximum values, with OpenCV.

Instead of using an image of a real transit and star, you'll draw circles on a black rectangle, just as you drew rectangles on the Mars map in the previous chapter. To plot the light curve, you can use matplotlib, Python's main plotting library. You installed matplotlib in "Installing NumPy and Other Scientific Packages with pip" on page 8 and began using it to make graphs in Chapter 2.

The Transit Code

The *transit.py* program uses OpenCV to generate a visual simulation of an exoplanet transiting a star, plots the resulting light curve with `matplotlib`, and estimates the size of the planet using the planetary radius equation from page 179. You can enter the code yourself or download it from *https://nostarch.com/real-world-python/*.

Importing Modules and Assigning Constants

Listing 8-1 imports modules and assigns constants representing user input.

transit.py, part 1

```
import math
import numpy as np
import cv2 as cv
import matplotlib.pyplot as plt

IMG_HT = 400
IMG_WIDTH = 500
BLACK_IMG = np.zeros((IMG_HT, IMG_WIDTH, 1), dtype='uint8')
STAR_RADIUS = 165
EXO_RADIUS = 7
EXO_DX = 3
EXO_START_X = 40
EXO_START_Y = 230
NUM_FRAMES = 145
```

Listing 8-1: Importing modules and assigning constants

Import the `math` module for the planetary radius equation, `NumPy` for calculating the brightness of the image, OpenCV for drawing the simulation, and `matplotlib` for plotting the light curve. Then start assigning constants that will represent user-input values.

Start with a height and width for the simulation window. The window will be a black, rectangular image built using the `np.zeros()` method, which returns an array of a given shape and type filled with zeros.

Recall that OpenCV images are `NumPy` arrays and items in the arrays must have the same type. The `uint8` data type represents an unsigned integer from 0 to 255. You can find a useful listing of other data types and their descriptions at *https://numpy.org/devdocs/user/basics.types.html*.

Next, assign radius values, in pixels, for the star and exoplanet. OpenCV will use these constants when it draws circles representing them.

The exoplanet will move across the face of the star, so you need to define how quickly it will move. The `EXO_DX` constant will increment the exoplanet's *x* position by three pixels with each programming loop, causing the exoplanet to move left to right.

Assign two constants to set the exoplanet's starting position. Then assign a `NUM_FRAMES` constant to control the number of simulation updates. Although you can calculate this number (`IMG_WIDTH/EXO_DX`), assigning it lets you fine-tune the duration of the simulation.

Defining the main() Function

Listing 8-2 defines the `main()` function used to run the program. Although you can define `main()` anywhere, placing it at the start lets it serve as a summary for the whole program, thus giving context to the functions defined later. As part of `main()`, you'll calculate the exoplanet's radius, nesting the equation within the call to the `print()` function.

transit.py, part 2

```
def main():
    intensity_samples = record_transit(EXO_START_X, EXO_START_Y)
    relative_brightness = calc_rel_brightness(intensity_samples)
    print('\nestimated exoplanet radius = {:.2f}\n'
          .format(STAR_RADIUS * math.sqrt(max(relative_brightness)
                                    - min(relative_brightness))))
    plot_light_curve(relative_brightness)
```

Listing 8-2: Defining the main() function

After defining the `main()` function, name a variable `intensity_samples` and call the `record_transit()` function. *Intensity* refers to the amount of light, represented by the numerical value of a pixel. The `record_transit()` function draws the simulation to the screen, measures its intensity, appends the measurement to a list called `intensity_samples`, and returns the list. It needs the starting point (*x*, *y*) coordinates for the exoplanet. Pass it the starting constants `EXO_START_X` and `EXO_START_Y`, which will place the planet in a position similar to ❶ in Figure 8-2. Note that if you increase the exoplanet's radius significantly, you may need to move the starting point farther to the left (negative values are acceptable).

Next, name a variable `relative_brightness` and call the `calc_rel_brightness()` function. As its name suggests, this function calculates *relative* brightness, which is the measured intensity divided by the maximum recorded intensity. It takes the list of intensity measurements as an argument, converts the measurements to relative brightness, and returns the new list.

You'll use the list of relative brightness values to calculate the radius of the exoplanet, in pixels, using the equation from page 179. You can perform the calculation as part of the `print()` function. Use the `{:.2f}` format to report the answer to two decimal points.

End the `main()` function by calling the function to plot the light curve. Pass it the `relative_brightness` list.

Recording the Transit

Listing 8-3 defines a function to simulate and record the transit event. It draws the star and exoplanet on a black rectangular image and then moves the exoplanet. It also calculates and displays the average intensity of the image with each move, appends the intensity to a list, and returns the list at the end.

```
def record_transit(exo_x, exo_y):
    """Draw planet transiting star and return list of intensity changes."""
    intensity_samples = []
    for _ in range(NUM_FRAMES):
        temp_img = BLACK_IMG.copy()
        cv.circle(temp_img, (int(IMG_WIDTH / 2), int(IMG_HT / 2)),
                STAR_RADIUS, 255, -1)
❶      cv.circle(temp_img, (exo_x, exo_y), EXO_RADIUS, 0, -1)
        intensity = temp_img.mean()
        cv.putText(temp_img, 'Mean Intensity = {}'.format(intensity), (5, 390),
                cv.FONT_HERSHEY_PLAIN, 1, 255)
        cv.imshow('Transit', temp_img)
        cv.waitKey(30)
❷      intensity_samples.append(intensity)
        exo_x += EXO_DX
    return intensity_samples
```

Listing 8-3: Drawing the simulation, calculating the image intensity, and returning it as a list

The record_transit() function takes a pair of (*x*, *y*) coordinates as arguments. These represent the starting point for the exoplanet or, more specifically, the pixel to use as the center of the first circle drawn in the simulation. It should not overlap with the star's circle, which will be centered in the image.

Next, create an empty list to hold the intensity measurements. Then start a for loop that uses the NUM_FRAMES constant to repeat the simulation a certain number of times. The simulation should last slightly longer than it takes for the exoplanet to exit the face of the star. That way, you get a full light curve that includes post-transit measurements.

Drawings and text placed on an image with OpenCV become part of that image. Consequently, you need to replace the previous image with each loop by copying the original BLACK_IMG to a local variable called temp_img.

Now you can draw the star by using the OpenCV circle() method. Pass it the temporary image, the (*x*, *y*) coordinates for the center of the circle that correspond to the center of the image, the STAR_RADIUS constant, a fill color of white, and a line thickness. Using a negative number for thickness fills the circle with color.

Draw the exoplanet circle next. Use the exo_x and exo_y coordinates as its starting point, the EXO_RADIUS constant as its size, and a black fill color ❶.

At this point, you should record the intensity of the image. Since the pixels already represent intensity, all you need to do is take the mean of the image. The number of measurements you take is dependent on the EXO_DX constant. The larger this value, the faster the exoplanet will move, and the fewer times you will record the mean intensity.

Display the intensity reading on the image using OpenCV's putText() method. Pass it the temporary image, a text string that includes the measurement, the (*x*, *y*) coordinates for the bottom-left corner of the text string, a font, a text size, and a color.

Now, name the window Transit and display it using OpenCV's imshow() method. Figure 8-5 shows a loop iteration.

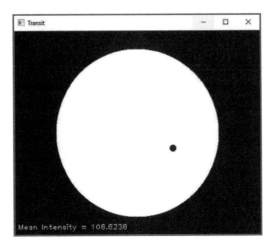

Figure 8-5: The exoplanet transiting the star

After showing the image, use the OpenCV waitKey() method to update it every 30 milliseconds. The lower the number passed to waitKey(), the faster the exoplanet will move across the star.

Append the intensity measurement to the intensity_samples list and then advance the exoplanet circle by incrementing its exo_x value by the EXO_DX constant ❷. Finish the function by returning the list of mean intensity measurements.

Calculating Relative Brightness and Plotting the Light Curve

Listing 8-4 defines functions to calculate the relative brightness of each intensity sample and display the light curve graph. It then calls the main() function if the program is not being used as a module in another program.

transit.py, part 4

```
def calc_rel_brightness(intensity_samples):
    """Return list of relative brightness from list of intensity values."""
    rel_brightness = []
    max_brightness = max(intensity_samples)
    for intensity in intensity_samples:
        rel_brightness.append(intensity / max_brightness)
    return rel_brightness

❶ def plot_light_curve(rel_brightness):
    """Plot changes in relative brightness vs. time."""
    plt.plot(rel_brightness, color='red', linestyle='dashed',
            linewidth=2, label='Relative Brightness')
    plt.legend(loc='upper center')
    plt.title('Relative Brightness vs. Time')
    plt.show()

❷ if __name__ == '__main__':
    main()
```

Listing 8-4: Calculating relative brightness, plotting the light curve, and calling main()

Light curves display the *relative* brightness over time so that an un-obscured star has a value of 1.0 and a totally eclipsed star has a value of 0.0. To convert the mean intensity measurements to relative values, define the calc_rel_brightness() function, which takes a list of mean intensity measurements as an argument.

Within the function, start an empty list to hold the converted values and then use Python's built-in max() function to find the maximum value in the intensity_samples list. To get relative brightness, loop through the items in this list and divide them by the maximum value. Append the result to the rel_brightness list as you go. End the function by returning the new list.

Define a second function to plot the light curve and pass it the rel _brightness list ❶. Use the matplotlib plot() method and pass it the list, a line color, a line style, a line width, and a label for the plot legend. Add the legend and plot title and then show the plot. You should see the chart in Figure 8-6.

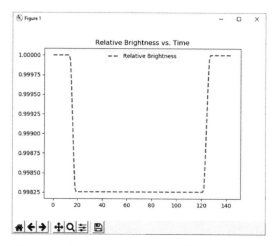

Figure 8-6: Example light curve plot from transit.py

The brightness variation on the plot might seem extreme at first glance, but if you look closely at the *y*-axis, you'll see that the exoplanet diminished the star's brightness by only 0.175 percent! To see how this looks on a plot of the star's absolute brightness (Figure 8-7), add the following line just before plt.show():

```
plt.ylim(0, 1.2)
```

The deflection in the light curve caused by the transit is subtle but detectable. Still, you don't want to go blind squinting at a light curve, so continue to let matplotlib automatically fit the *y*-axis as in Figure 8-6.

Finish the program by calling the main() function ❷. In addition to the light curve, you should see the estimated radius of the exoplanet in the shell.

```
estimated exoplanet radius = 6.89
```

That's all there is to it. With fewer than 50 lines of Python code, you've developed a means of discovering exoplanets!

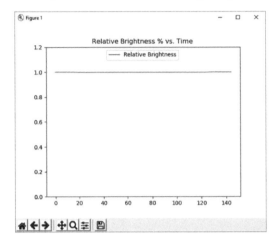

Figure 8-7: Light curve from Figure 8-6 with rescaled y-axis

Experimenting with Transit Photometry

Now that you have a working simulation, you can use it to model the behavior of transits, allowing you to better analyze real-life observations you'll make in the future. One approach would be to run a lot of possible cases and produce an "atlas" of expected exoplanet responses. Researchers could use this atlas to help them interpret actual light curves.

For example, what if the plane of an exoplanet's orbit is tilted with respect to Earth so that the exoplanet only partly crosses the star during transit? Would researchers be able to detect its position from its light curve signature, or would it just look like a smaller exoplanet doing a complete transit?

If you run the simulation with an exoplanet radius of 7 and let it skim the base of the star, you should get a U-shaped curve (Figure 8-8).

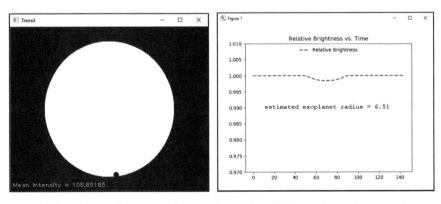

Figure 8-8: Light curve for an exoplanet with a radius of 7 that only partly crosses its star

If you run the simulation again with an exoplanet radius of 5 and let the exoplanet pass fully over the face of the star, you get the graph in Figure 8-9.

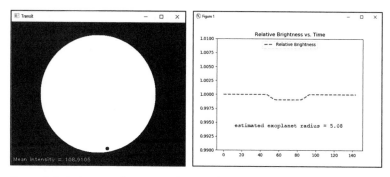

Figure 8-9: Light curve for an exoplanet with a radius of 5 that fully crosses its star

When an exoplanet skims the side of a star, never fully passing over it, the overlapping area changes constantly, generating the U-shaped curve in Figure 8-8. If the entire exoplanet passes over the face of the star, the base of the curve is flatter, as in Figure 8-9. And because you never see the planet's full disc against the star in a partial transit, you have no way to measure its true size. Thus, size estimates should be taken with a grain of salt if your light curve lacks a flattish bottom.

If you run a range of exoplanet sizes, you'll see that the light curve changes in predictable ways. As size increases, the curve deepens, with longer ramps on either side, because a larger fraction of the star's brightness is diminished (Figures 8-10 and 8-11).

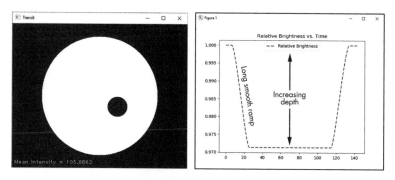

Figure 8-10: Light curve for EXO_RADIUS = 28

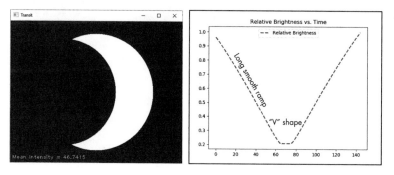

Figure 8-11: Light curve for EXO_RADIUS = 145

Because exoplanets are circular objects with smooth edges, they should produce light curves with smooth ramps that continuously increase or decrease. This is important knowledge, as astronomers have recorded decidedly bumpy curves when looking for exoplanets. In the "Practice Project" sections at the end of the chapter, you'll use your program to explore oddly shaped light curves that might be explained by extraterrestrial engineering!

Project #12: Imaging Exoplanets

By 2025, three powerful telescopes—two on Earth and one in space—will use infrared and visible light to directly image Earth-sized exoplanets. In the best-case scenario, the exoplanet will show up as a single saturated pixel with some bleed into the surrounding pixels, but that's enough to tell whether the planet rotates, has continents and seas, experiences weather and seasons, and could support life as we know it!

In this project, you'll simulate the process of analyzing an image taken from those telescopes. You'll use Earth as a stand-in for a distant exoplanet. This way, you can easily relate known features, such as continents and oceans, to what you see in a single pixel. You'll focus on the color composition and intensity of reflected light and make inferences about the exoplanet's atmosphere, surface features, and rotation.

THE OBJECTIVE

Write a Python program that pixelates images of Earth and plots the intensity of the red, green, and blue color channels.

The Strategy

To demonstrate that you can capture different surface features and cloud formations with a single saturated pixel, you need only two images: one of the western hemisphere and one of the eastern. Conveniently, NASA has already photographed both hemispheres of Earth from space (Figure 8-12).

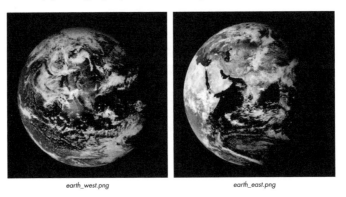

earth_west.png *earth_east.png*

Figure 8-12: Images of the western and eastern hemispheres

The size of these images is 474×474 pixels, a resolution far too high for a future exoplanet image, where the exoplanet is expected to occupy 9 pixels, with only the center pixel fully covered by the planet (Figure 8-13).

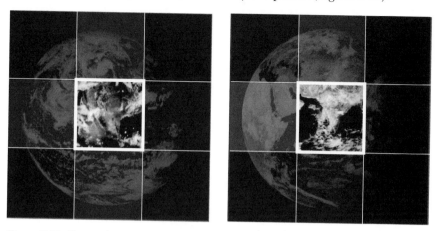

Figure 8-13: The earth_west.png and earth_east.png images overlaid with a 9-pixel grid

You'll need to degrade the Earth images by mapping them into a 3×3 array. Since OpenCV uses NumPy, this will be easy to do. To detect changes in the exoplanet's surface, you'll need to extract the dominant colors (blue, green, and red). OpenCV will let you average these color channels. Then you can display the results with matplotlib.

The Pixelator Code

The *pixelator.py* program loads the two images of Earth, resizes them to 3×3 pixels, and then resizes them again to 300×300 pixels. These final images are just for visualization; they have the same color information as the 3×3 images. The program then averages the color channels in both resized images and plots the results as pie charts that you can compare. You can download the code and two images (*earth_west.png* and *earth_east.png*) from the book's website. Keep them in the same folder and don't rename the images.

Importing Modules and Downscaling Images

Listing 8-5 imports modules for plotting and image processing and then loads and degrades two images of Earth. It first reduces each to 9 pixels in a 3×3 array. It then enlarges the decimated images to 300×300 pixels so they are large enough to see and posts them to the screen.

pixelator.py, part 1

```
import numpy as np
import cv2 as cv
from matplotlib import pyplot as plt

files = ['earth_west.png', 'earth_east.png']

for file in files:
```

```
img_ini = cv.imread(file)
pixelated = cv.resize(img_ini, (3, 3), interpolation=cv.INTER_AREA)
img = cv.resize(pixelated, (300, 300), interpolation=cv.INTER_AREA)
cv.imshow('Pixelated {}'.format(file), img)
cv.waitKey(2000)
```

Listing 8-5: Importing modules and loading, degrading, and showing images

Import NumPy and OpenCV to work with the images and use matplotlib to plot their color components as pie charts. Then start a list of filenames containing the two images of Earth.

Now start looping through the files in the list and use OpenCV to load them as NumPy arrays. Recall that OpenCV loads color images by default, so you don't need to add an argument for this.

Your goal is to reduce the image of Earth into a single saturated pixel surrounded by partially saturated pixels. To degrade the images from their original 474×474 size to 3×3, use OpenCV's resize() method. First, name the new image *pixelated* and pass the method the current image, the new width and height in pixels, and an interpolation method. *Interpolation* occurs when you resize an image and use known data to estimate values at unknown points. The OpenCV documentation recommends the INTER_AREA interpolation method for shrinking images (see the geometric image transformations at *https://docs.opencv.org/4.3.0/da/d54/group__imgproc__transform.html*).

At this point, you have a tiny image that's too small to visualize, so resize it again to 300×300 so you can check the results. Use either INTER_NEAREST or INTER_AREA as the interpolation method, as these will preserve the pixel boundaries.

Show the image (Figure 8-14) and delay the program for two seconds using waitKey().

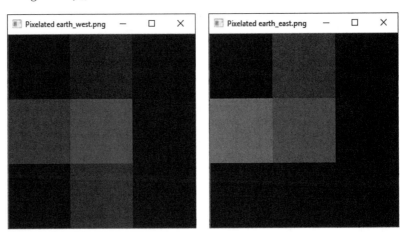

Figure 8-14: Grayscale view of the pixelated color images

Note that you can't restore the images to their original state by resizing them to 474×474. Once you average the pixel values down to a 3×3 matrix, all the detailed information is lost forever.

Averaging the Color Channels and Making the Pie Charts

Still in the for loop, Listing 8-6 makes and displays pie charts of the blue, green, and red color components of each pixelated image. You can compare these to make inferences about the planet's weather, landmasses, rotation, and so on.

pixelator.py, part 2

```
b, g, r = cv.split(pixelated)
color_aves = []
for array in (b, g, r):
    color_aves.append(np.average(array))

labels = 'Blue', 'Green', 'Red'
colors = ['blue', 'green', 'red']
fig, ax = plt.subplots(figsize=(3.5, 3.3))  # size in inches
❶ _, _, autotexts = ax.pie(color_aves,
                           labels=labels,
                           autopct='%1.1f%%',
                           colors=colors)
for autotext in autotexts:
    autotext.set_color('white')
plt.title('{}\n'.format(file))

plt.show()
```

Listing 8-6: Splitting out and averaging color channels and making a pie chart of colors

Use OpenCV's split() method to break out the blue, green, and red color channels in the pixelated image and unpack the results into b, g, and r variables. These are arrays, and if you call print(b), you should see this output:

```
[[ 49  93  22]
 [124 108  65]
 [ 52 118  41]]
```

Each number represents a pixel—specifically, the pixel's blue value—in the 3×3 pixelated image. To average the arrays, first make an empty list to hold the averages and then loop through the arrays and call the NumPy average method, appending the results to the list.

Now you're ready to make pie charts of the color averages in each pixelated image. Start by assigning color names to a variable named labels, which you'll use to annotate the pie wedges. Next, specify the colors you want to use in the pie chart. These will override the matplotlib default choices. To make the chart, use the fig, ax naming convention for figure and axis, call the subplots() method, and pass it a figure size in inches.

Because the colors will vary only slightly between images, you'll want to post the percentage of each color in its pie wedge so you can easily see whether there's a difference between them. Unfortunately, the matplotlib default is to use black text that can be hard to see against a dark background. To fix this, call the ax.pie() method for making pie charts and use its autotexts list ❶. The method returns three lists, one concerning the pie wedges, one concerning the labels, and one for numeric labels, called *autotexts*. You need only the last one, so treat the first two as unused variables by assigning them to an underscore symbol.

Pass ax.pie() the list of color averages and the list of labels and set its autopct parameter to show numbers to one decimal place. If this parameter is set to None, the autotexts list will not be returned. Finish the arguments by passing the list of colors to use for the pie wedges.

The autotexts list for the first image looks like this:

```
[Text(0.18326840031431146, 0.5713253822554821, '40.1%'), Text(-0.5646237442340427,
-0.20297789891298565, '30.7%'), Text(0.36574010704848686, -0.47564080364930983, '29.1%')
```

Each Text object has (*x*, *y*) coordinates and a percent value as a text string. These will still post in black, so you need to loop through the objects and change the color to white using their set_color() method. Now all you need to do is set the chart title to the filename and show the plots (Figure 8-15).

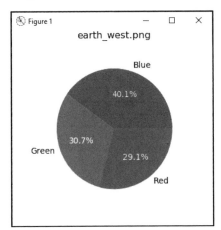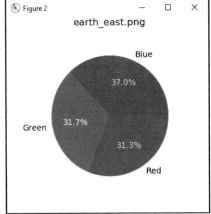

Figure 8-15: The pie charts produced by pixelator.py

Although the pie charts are similar, the differences are meaningful. If you compare the original color images, you'll see that the *earth_west.png* photograph includes more ocean and should produce a larger blue wedge.

Plotting a Single Pixel

The charts in Figure 8-15 are for the whole image, which includes a sampling of black space. For an uncontaminated sample, you could use the single saturated pixel at the center of each image, as shown in Listing 8-7.

This code represents an edited copy of *pixelator.py*, with the lines that change annotated. You can find a digital copy in the *Chapter_8* folder as *pixelator_saturated_only.py*.

pixelator_ saturated_only.py

```
import cv2 as cv
from matplotlib import pyplot as plt

files = ['earth_west.png', 'earth_east.png']

# Downscale image to 3x3 pixels.
for file in files:
    img_ini = cv.imread(file)
    pixelated = cv.resize(img_ini, (3, 3), interpolation=cv.INTER_AREA)
    img = cv.resize(pixelated, (300, 300), interpolation=cv.INTER_NEAREST)
    cv.imshow('Pixelated {}'.format(file), img)
    cv.waitKey(2000)

❶  color_values = pixelated[1, 1]  # Selects center pixel.

    # Make pie charts.
    labels = 'Blue', 'Green', 'Red'
    colors = ['blue', 'green', 'red']
    fig, ax = plt.subplots(figsize=(3.5, 3.3))  # Size in inches.
❷  _, _, autotexts = ax.pie(color_values,
                            labels=labels,
                            autopct='%1.1f%%',
                            colors=colors)
    for autotext in autotexts:
        autotext.set_color('white')
❸  plt.title('{} Saturated Center Pixel \n'.format(file))

plt.show()
```

Listing 8-7: Plotting pie charts for the colors in the center pixel of the pixelated image

The four lines of code in Listing 8-6 that split the image and averaged the color channels can be replaced with one line ❶. The pixelated variable is a NumPy array, and [1, 1] represents row 1, column 1 in the array. Remember that Python starts counting at 0, so these values correspond to the center of a 3×3 array. If you print the color_values variable, you'll see another array.

```
[108 109 109]
```

These are the blue, green, and red color channel values for the center pixel, and you can pass them directly to matplotlib ❷. For clarity, change the plot title so it indicates that you're analyzing the center pixel only ❸. Figure 8-16 shows the resulting plots.

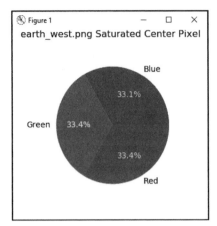
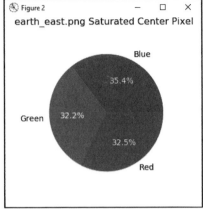

Figure 8-16: The single-pixel pie charts produced by pixelator_saturated_only.py

The color differences between the western and eastern hemispheres in Figures 8-15 and 8-16 are subtle, but you know they're real because you *forward modeled* the response. That is, you produced the result from actual observations, so you know the result is meaningful, repeatable, and unique.

In a real exoplanet survey, you'd want to take as many images as possible. If similar intensity and color patterns persist over time, then you can rule out stochastic effects such as weather. If the color patterns change predictably over long time periods, you may be seeing the effect of seasons, such as the presence of white polar caps in the winter and the spread of green vegetation in the spring and summer.

If measurements repeat periodically over relatively short time spans, you can infer that the planet is rotating on its axis. In the "Practice Project" sections at the end of the chapter, you'll get a chance to calculate the length of an exoplanet's day.

Summary

In this chapter, you used OpenCV, `NumPy`, and `matplotlib` to create images and measure their properties. You also resized images to different resolutions and plotted image intensity and color channel information. With short and simple Python programs, you simulated important methods that astronomers use to discover and study distant exoplanets.

Further Reading

How to Search for Exoplanets, by the Planetary Society (*https://www.planetary.org/*), is a good overview of the techniques used to search for exoplanets, including the strengths and weaknesses of each method.

"Transit Light Curve Tutorial," by Andrew Vanderburg, explains the basics of the transit photometry method and provides links to Kepler Space Observatory transit data. You can find it at *https://www.cfa.harvard.edu /~avanderb/tutorial/tutorial.html*.

"NASA Wants to Photograph the Surface of an Exoplanet" (Wired, 2020), by Daniel Oberhaus, describes the effort to turn the sun into a giant camera lens for studying exoplanets.

"Dyson Spheres: How Advanced Alien Civilizations Would Conquer the Galaxy" (Space.com, 2014), by Karl Tate, is an infographic on how an advanced civilization could capture the power of a star using vast arrays of solar panels.

Ringworld (Ballantine Books, 1970), by Larry Niven, is one of the classic novels of science fiction. It tells the story of a mission to a massive abandoned alien construct—the Ringworld—that encircles an alien star.

Practice Project: Detecting Alien Megastructures

In 2015, citizen scientists working on data from the Kepler space telescope noticed something odd about Tabby's Star, located in the constellation Cygnus. The star's light curve, recorded in 2013, exhibited irregular changes in brightness that were far too large to be caused by a planet (Figure 8-17).

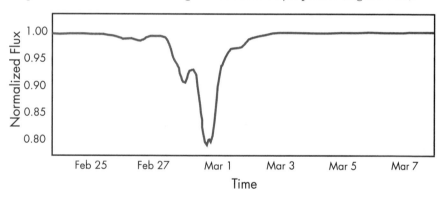

Figure 8-17: Light curve for Tabby's Star, measured by the Kepler Space Observatory

Besides the dramatic drop in brightness, the light curve was asymmetrical and included weird bumps that aren't seen in typical planetary transits. Proposed explanations posited that the light curve was caused by the consumption of a planet by the star, the transit of a cloud of disintegrating comets, a large ringed planet trailed by swarms of asteroids, or an *alien megastructure*.

Scientists speculated that an artificial structure of this size was most likely an attempt by an alien civilization to collect energy from its sun. Both science literature and science fiction describe these staggeringly large solar panel projects. Examples include Dyson swarms, Dyson spheres, ringworlds, and Pokrovsky shells (Figure 8-18).

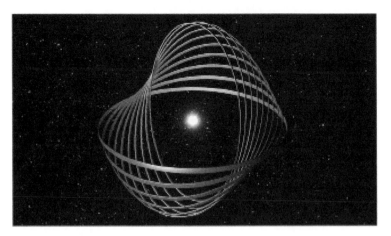

Figure 8-18: Pokrovsky shell system of rings around a star designed to intercept the star's radiation

In this practice project, use the *transit.py* program to approximate the shape and depth of the Tabby's Star light curve. Replace the circular exoplanet used in the program with other simple geometric shapes. You don't need to match the curve exactly; just capture key features such as the asymmetry, the "bump" seen around February 28, and the large drop in brightness.

You can find my attempt, *practice_tabbys_star.py*, in the *Chapter_8* folder, downloadable from the book's website at *https://nostarch.com/real-world-python/*, and in the appendix. It produces the light curve shown in Figure 8-19.

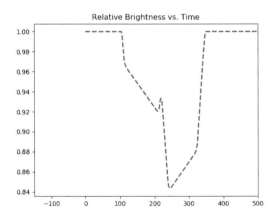

Figure 8-19: Light curve produced by practice_tabbys_star.py

We now know that whatever is orbiting Tabby's Star allows some wavelengths of light to pass, so it can't be a solid object. Based on this behavior and the wavelengths it absorbed, scientists believe dust is responsible for the weird shape of the star's light curve. Other stars, however, like HD 139139 in the constellation Libra, have bizarre light curves that remain unexplained at the time of this writing.

Practice Project: Detecting Asteroid Transits

Asteroid fields may be responsible for some bumpy and asymmetrical light curves. These belts of debris often originate from planetary collisions or the creation of a solar system, like the Trojan asteroids in Jupiter's orbit (Figure 8-20). You can find an interesting animation of the Trojan asteroids on the web page "Lucy: The First Mission to the Trojan Asteroids" at *https://www.nasa.gov/*.

Figure 8-20: More than one million Trojan asteroids share Jupiter's orbit.

Modify the *transit.py* program so that it randomly creates asteroids with radii between 1 and 3, weighted heavily toward 1. Allow the user to input the number of asteroids. Don't bother calculating the exoplanet radius, since the calculation assumes you're dealing with a single spherical object, which you're not. Experiment with the number of asteroids, the size of the asteroids, and the spread (the *x*-range and *y*-range in which the asteroids exist) to see the impact on the light curve. Figure 8-21 shows one such example.

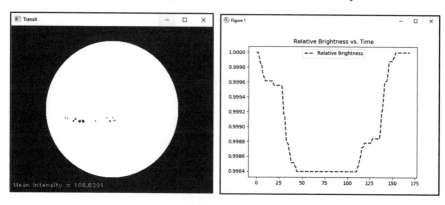

Figure 8-21: Irregular, asymmetrical light curve produced by a randomly generated asteroid field

You can find a solution, *practice_asteroids.py*, in the appendix and on the book's web page. This program uses object-oriented programming (OOP) to simplify the management of multiple asteroids.

Practice Project: Incorporating Limb Darkening

The *photosphere* is the luminous outer layer of a star that radiates light and heat. Because the temperature of the photosphere falls as the distance from the star's center increases, the edges of a star's disk are cooler and therefore appear dimmer than the center of the star (Figure 8-22). This effect is known as *limb darkening*.

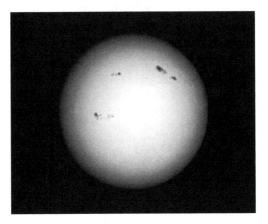

Figure 8-22: Limb darkening and sunspots on the sun

Rewrite the *transit.py* program so that it addresses limb darkening. Rather than draw the star, use the image *limb_darkening.png* in the *Chapter_8* folder, downloadable from the book's website.

Limb darkening will affect the light curves produced by planetary transits. Compared to the theoretical curves you produced in Project 11, they will appear less boxy, with rounder, softer edges and a curved bottom (Figure 8-23).

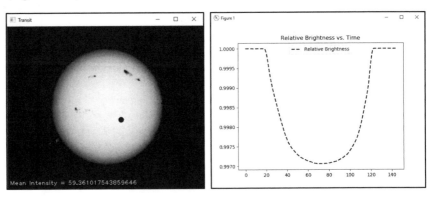

Figure 8-23: The effect of limb darkening on a light curve

Use your modified program to revisit "Experimenting with Transit Photometry" on page 186, where you analyzed the light curves produced by partial transits. You should see that, compared to partial transits, full transits still produce broader dips with flattish bottoms (Figure 8-24).

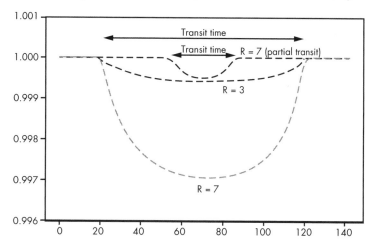

Figure 8-24: Limb-darkened light curves for full and partial transits (R = exoplanet radius)

If the full transit of a small planet occurs near the edge of a star, limb darkening may make it difficult to distinguish from the partial transit of a larger planet. You can see this in Figure 8-25, where arrows denote the location of the planets.

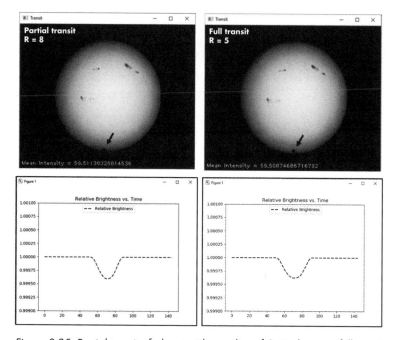

Figure 8-25: Partial transit of planet with a radius of 8 pixels versus full transit of planet with a radius of 5 pixels

Astronomers have many tools for extracting information entangled in a light curve. By recording multiple transit events, they can determine an exoplanet's orbital parameters, such as the distance between the planet and the star. They can use subtle inflections in the light curve to tease out the amount of time the planet is fully over the surface of the star. They can estimate the theoretical amount of limb darkening, and they can use modeling, as you're doing here, to bring it all together and test their assumptions against actual observations.

You can find a solution, *practice_limb_darkening.py*, in the appendix and in the *Chapter_8* folder downloadable from the book's website.

Practice Project: Detecting Starspots

Sunspots—called *starspots* on alien suns—are regions of reduced surface temperature caused by variations in the star's magnetic field. Starspots can darken the face of stars and do interesting things to light curves. In Figure 8-26, an exoplanet passes over a starspot, causing a "bump" in the light curve.

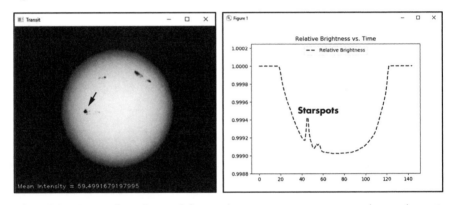

Figure 8-26: An exoplanet (arrow, left image) passing over a starspot produces a bump in the light curve.

To experiment with starspots, use the *practice_limb_darkening.py* code from the previous practice project and edit it so that an exoplanet roughly the same size as the starspots passes over them during its transit. To reproduce Figure 8-26, use EXO_RADIUS = 4, EXO_DX = 3, and EXO_START_Y = 205.

Practice Project: Detecting an Alien Armada

The hyper-evolved beavers of exoplanet BR549 have been as busy as, well, beavers. They've amassed an armada of colossal colony ships that are now loaded and ready to leave orbit. Thanks to some exoplanet detection of their own, they've decided to abandon their chewed-out homeworld for the lush green forests of Earth!

Write a Python program that simulates multiple spaceships transiting a star. Give the ships different sizes, shapes, and speeds (such as those in Figure 8-27).

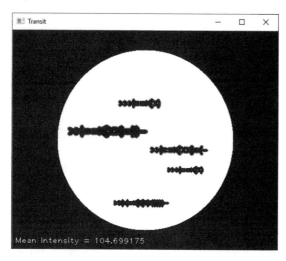

Figure 8-27: An armada of alien colony ships preparing to invade Earth

Compare the resultant light curves to those from Tabby's Star (Figure 8-17) and the asteroids practice project. Do the ships produce distinctive curves, or can you get similar patterns from asteroid swarms, starspots, or other natural phenomena?

You can find a solution, *practice_alien_armada.py*, in the appendix and in the *Chapter_8* folder, downloadable from the book's website.

Practice Project: Detecting a Planet with a Moon

What kind of light curve would an exoplanet with an orbiting moon produce? Write a Python program that simulates a small exomoon orbiting a larger exoplanet and calculate the resulting light curve. You can find a solution, *practice_planet_moon.py*, in the appendix and on the book's website.

Practice Project: Measuring the Length of an Exoplanet's Day

Your astronomer boss has given you 34 images of an exoplanet designated BR549. The images were taken an hour apart. Write a Python program that loads the images in order, measures the intensity of each image, and plots the measurements as a single light curve (Figure 8-28). Use the curve to determine the length of a day on BR549.

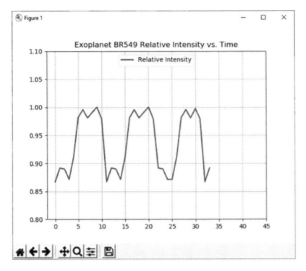

Figure 8-28: Composite light curve for 34 images of exoplanet BR549

You can find a solution, *practice_length_of_day.py*, in the appendix. The digital version of the code, along with the folder of images (*br549_pixelated*), are in the *Chapter_8* folder downloadable from the book's website.

Challenge Project: Generating a Dynamic Light Curve

Rewrite *transit.py* so that the light curve dynamically updates as the simulation runs, rather than just appearing at the end.

9

IDENTIFYING FRIEND OR FOE

Face detection is a machine learning technology that locates human faces in digital images. It's the first step in the process of face *recognition*, a technique for identifying individual faces using code. Face detection and recognition methods have broad applications, such as tagging photographs on social media, autofocusing digital cameras, unlocking cell phones, finding missing children, tracking terrorists, facilitating secure payments, and more.

In this chapter, you'll use machine learning algorithms in OpenCV to program a robot sentry gun. Because you'll be distinguishing between humans and otherworldly mutants, you'll only need to detect the *presence* of human faces rather than identify specific individuals. In Chapter 10, you'll take the next step and identify people by their faces.

Detecting Faces in Photographs

Face detection is possible because human faces share similar patterns. Some common facial patterns are the eyes being darker than the cheeks and the bridge of the nose being brighter than the eyes, as seen in the left image of Figure 9-1.

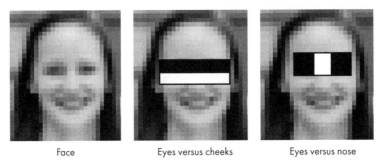

Face   Eyes versus cheeks   Eyes versus nose

Figure 9-1: Example of some consistently bright and dark regions in a face

You can extract these patterns using templates like those in Figure 9-2. These yield *Haar features*, a fancy name for the attributes of digital images used in object recognition. To calculate a Haar feature, place one of the templates on a grayscale image, add up the grayscale pixels that overlap with the white part, and subtract them from the sum of the pixels that overlap the black part. Thus, each feature consists of a single intensity value. We can use a range of template sizes to sample all possible locations on the image, making the system scale invariant.

Figure 9-2: Some example Haar feature templates

In the middle image in Figure 9-1, an "edge feature" template extracts the relationship between the dark eyes and the bright cheeks. In the far-right image in Figure 9-1, a "line feature" template extracts the relationship between the dark eyes and the bright nose.

By calculating Haar features on thousands of *known* face and nonface images, we can determine which combination of Haar features is most effective at identifying faces. This training process is slow, but it facilitates fast detection later. The resulting algorithm, known as a *face classifier*, takes the values of features in an image and predicts whether it contains a human face by outputting 1 or 0. OpenCV ships with a pretrained face detection classifier based on this technique.

To apply the classifier, the algorithm uses a *sliding window* approach. A small rectangular area is incrementally moved across the image and evaluated using a *cascade classifier* consisting of multiple stages of filters. The filters at each stage are combinations of Haar features. If the window region fails to pass the threshold of a stage, it's rejected, and the window slides to the next position. Quickly rejecting nonface regions, like the one shown in the right inset in Figure 9-3, helps speed up the overall process.

Figure 9-3: Images are searched for faces using a rectangular sliding window.

If a region passes the threshold for a stage, the algorithm processes another set of Haar features and compares them to the threshold, and so on, until it either rejects or positively identifies a face. This causes the sliding window to speed up or slow down as it moves across an image. You can find a fantastic video example of this at *https://vimeo.com/12774628/*.

For each face detected, the algorithm returns the coordinates of a rectangle around the face. You can use these rectangles as the basis for further analysis, such as identifying eyes.

Project #13: Programming a Robot Sentry Gun

Imagine that you're a technician with the Coalition Marines, a branch of the Space Force. Your squad has been deployed to a secret research base operated by the Wykham-Yutasaki Corporation on planet LV-666. While studying a mysterious alien apparatus, the researchers have inadvertently opened a portal to a hellish alternate dimension. Anyone who gets near the portal, including dozens of civilians and several of your comrades, mutates into a murderous mindless monstrosity! You've even caught security footage of the result (Figure 9-4).

Figure 9-4: Security camera footage of a mutated scientist (left) and marine (right)

According to the remaining scientists, the mutation affects more than just organic matter. Any gear the victim is wearing, such as helmets and goggles, is also transmogrified and fused into the flesh. Eye tissue is especially vulnerable. All the mutants formed so far are eyeless and blind, though this doesn't seem to affect their mobility. They're still ferocious, deadly, and unstoppable without military-grade weapons.

That's where you come in. It's your job to set up an automatic firing station to guard Corridor 5, a key access point in the compromised facility. Without it, your small squad is in danger of being outflanked and overrun by hordes of rampaging mutants.

The firing station consists of a UAC 549-B automated sentry gun, called a *robot sentry* by the grunts (Figure 9-5). It's equipped with four M30 autocannons with 1,000 rounds of ammo and multiple sensors, including a motion detector, laser ranging unit, and optical camera. The gun also interrogates targets using an identification friend or foe (IFF) transponder. All Coalition Marines carry these transponders, allowing them to safely pass active sentry guns.

Figure 9-5: A UAC 549-B automated sentry gun

Unfortunately, the squad's sentry gun was damaged during landing, so the transponders no longer function. Worse, the requisitions corporal forgot to download the software that visually interrogates targets. With the transponder sensor down, there's no way to positively identify marines and civilians. You'll need to get this fixed as quickly as possible, because your buddies are badly outnumbered and the mutants are on the move!

Fortunately, planet LV-666 has no indigenous life forms, so you need to distinguish between humans and mutants only. Since the mutants are basically faceless, a face detection algorithm is the logical solution.

THE OBJECTIVE

Write a Python program that disables the sentry gun's firing mechanism when it detects human faces in an image.

The Strategy

In situations like this, it's best to keep things simple and leverage existing resources. This means relying on OpenCV's face detection functionality rather than writing customized code to recognize the humans on the base. But you can't be sure how well these canned procedures will work, so you'll need to guide your human targets to make the job as easy as possible.

The sentry gun's motion detector will handle the job of triggering the optical identification process. To permit humans to pass unharmed, you'll need to warn them to stop and face the camera. They'll need a few seconds to do this and a few seconds to proceed past the gun after they're cleared.

You'll also want to run some tests to ensure OpenCV's training set is adequate and you're not generating any false positives that would let a mutant sneak by. You don't want to kill anyone with friendly fire, but you can't be too cautious, either. If one mutant gets by, everyone could perish.

NOTE *In real life, the sentry guns would use a video feed. Since I don't have my own film studio with special effects and makeup departments, you'll work off still photos instead. You can think of these as individual video frames. Later in the chapter, you'll get a chance to detect your own face using your computer's video camera.*

The Code

The *sentry.py* code will loop through a folder of images, identify human faces in the images, and show the image with the faces outlined. It will then either fire or disable the gun depending on the result. You'll use the images in the *corridor_5* folder in the *Chapter_9* folder, downloadable from *https://nostarch.com/real-world-python/*. As always, don't move or rename any files after downloading and launch *sentry.py* from the folder in which it's stored.

You'll also need to install two modules, playsound and pyttsx3. The first is a cross-platform module for playing WAV and MP3 format audio files.

You'll use it to produce sound effects, such as machine gun fire and an "all clear" tone. The second is a cross-platform wrapper that supports the native text-to-speech libraries on Windows and Linux-based systems, including macOS. The sentry gun will use this to issue audio warnings and instructions. Unlike other text-to-speech libraries, pyttsx3 reads text directly from the program, rather than first saving it to an audio file. It also works offline, making it reliable for voice-based projects.

You can install both modules with pip in a PowerShell or Terminal window.

```
pip install playsound
pip install pyttsx3
```

If you encounter an error installing pyttsx3 on Windows, such as No module named win32.com.client, No module named win32, or No module named win32api, then install pypiwin32.

```
pip install pypiwin32
```

You may need to restart the Python shell and editor following this installation.

For more on playsound, see *https://pypi.org/project/playsound/*. The documentation for pyttsx3 can be found at *https://pyttsx3.readthedocs.io/en/latest/* and *https://pypi.org/project/pyttsx3/*.

If you don't already have OpenCV installed, see "Installing the Python Libraries" on page 6.

Importing Modules, Setting Up Audio, and Referencing the Classifier Files and Corridor Images

Listing 9-1 imports modules, initializes and sets up the audio engine, assigns the classifier files to variables, and changes the directory to the folder containing the corridor images.

sentry.py, part 1

```
import os
import time
❶ from datetime import datetime
from playsound import playsound
import pyttsx3
import cv2 as cv

❷ engine = pyttsx3.init()
engine.setProperty('rate', 145)
engine.setProperty('volume', 1.0)

root_dir = os.path.abspath('.')
gunfire_path = os.path.join(root_dir, 'gunfire.wav')
tone_path = os.path.join(root_dir, 'tone.wav')

❸ path= "C:/Python372/Lib/site-packages/cv2/data/"
face_cascade = cv.CascadeClassifier(path +
                               'haarcascade_frontalface_default.xml')
```

```
    eye_cascade = cv.CascadeClassifier(path + 'haarcascade_eye.xml')

❹ os.chdir('corridor_5')
    contents = sorted(os.listdir())
```

Listing 9-1: Importing modules, setting up the audio, and locating the classifier files and corridor images

Except for datetime, playsound, and pytts3, the imports should be familiar to you if you've worked through the earlier chapters ❶. You'll use datetime to record the exact time at which an intruder is detected in the corridor.

To use pytts3, initialize a pyttsx3 object and assign it to a variable named, by convention, engine ❷. According to the pyttsx3 docs, an application uses the engine object to register and unregister event callbacks, produce and stop speech, get and set speech engine properties, and start and stop event loops.

In the next two lines, set the rate of speech and volume properties. The rate of speech value used here was obtained through trial and error. It should be fast but still clearly understandable. The volume should be set to the maximum value (1.0) so any humans stumbling into the corridor can easily hear the warning instructions.

The default voice on Windows is male, but other voices are available. For example, on a Windows 10 machine, you can switch to a female voice using the following voice ID:

```
engine.setProperty('voice',
'HKEY_LOCAL_MACHINE\SOFTWARE\Microsoft\Speech\Voices\Tokens\TTS_MS_EN-US_ZIRA_11.0')
```

To see a list of voices available on your platform, refer to "Changing voices" at *https://pyttsx3.readthedocs.io/en/latest/*.

Next, set up the audio recording of gunfire, which you'll play when a mutant is detected in the corridor. Specify the location of the audio file by generating a directory path string that will work on all platforms, which you do by combining the absolute path with the filename using the os.path.join() method. Use the same path for the *tone.wav* file, which you'll use as an "all clear" signal when the program identifies a human.

The pretrained Haar cascade classifiers should download as *.xml* files when you install OpenCV. Assign the path for the folder containing the classifiers to a variable ❸. The path shown is for my Windows machine; your path may be different. On macOS, for example, you may find them under *opencv/data/haarcascades*. You can also find them online at *https://github.com/opencv/opencv/tree/master/data/haarcascades/*.

Another option for finding the path to the cascade classifiers is to use the preinstalled sysconfig module, as in the following snippet:

```
>>> import sysconfig
>>> path = sysconfig.get_paths()['purelib'] + '/cv2/data'
>>> path
'C:\\Python372\\Lib\\site-packages/cv2/data'
```

This should work for Windows inside and outside of virtual environments. However, this will work on Ubuntu only within a virtual environment.

To load a classifier, use OpenCV's CascadeClassifier() method. Use string concatenation to add the path variable to the filename string for the classifier and assign the result to a variable.

Note that I use only two classifiers, one for frontal faces and one for eyes, to keep things simple. Additional classifiers are available for profiles, smiles, eyeglasses, upper bodies, and so on.

Finish by pointing the program to the images taken in the corridor you are guarding. Change the directory to the proper folder ❹; then list the folder contents and assign the results to a contents variable. Because you're not providing a full path to the folder, you'll need to launch your program from the folder containing it, which should be one level above the folder with the images.

Issuing a Warning, Loading Images, and Detecting Faces

Listing 9-2 starts a for loop to iterate through the folder containing the corridor images. In real life, the motion detectors on the sentry guns would launch your program as soon as something entered the corridor. Since we don't have any motion detectors, we'll assume that each loop represents the arrival of a new intruder.

The loop immediately arms the gun and prepares it to fire. It then verbally requests that the intruder stop and face the camera. This would occur at a set distance from the gun, as determined by the motion detector. As a result, you know the faces will all be roughly the same size, making it easy to test the program.

The intruder is given a few seconds to comply with the command. After that, the cascade classifier is called and used to search for faces.

sentry.py, part 2

```
for image in contents:
❶   print(f"\nMotion detected...{datetime.now()}")
    discharge_weapon = True
❷   engine.say("You have entered an active fire zone. \
                Stop and face the gun immediately. \
                When you hear the tone, you have 5 seconds to pass.")
    engine.runAndWait()
    time.sleep(3)

❸   img_gray = cv.imread(image, cv.IMREAD_GRAYSCALE)
    height, width = img_gray.shape
    cv.imshow(f'Motion detected {image}', img_gray)
    cv.waitKey(2000)
    cv.destroyWindow(f'Motion detected {image}')

❹   face_rect_list = []
    face_rect_list.append(face_cascade.detectMultiScale(image=img_gray,
                                            scaleFactor=1.1,
                                            minNeighbors=5))
```

Listing 9-2: Looping through images, issuing a verbal warning, and searching for faces

Start looping through the images in the folder. Each new image represents a new intruder in the corridor. Print a log of the event and the time at which it occurred ❶. Note the f before the start of the string. This is the new *f-string* format introduced with Python 3.6 (*https://www.python.org/dev/peps/pep-0498/*). An f-string is a literal string that contains expressions, such as variables, strings, mathematical operations, and even function calls, inside curly braces. When the program prints the string, it replaces the expressions with their values. These are the fastest and most efficient string formats in Python, and we certainly want this program to be fast!

Assume every intruder is a mutant and prepare to discharge the weapon. Then, verbally warn the intruder to stop and be scanned.

Use the pyttsx3 engine object's say() method to speak ❷. It takes a string as an argument. Follow this with the runAndWait() method. This halts program execution, flushes the say() queue, and plays the audio.

NOTE *For some macOS users, the program may exit with the second call to runAndWait(). If this occurs, download the* sentry_for_Mac_bug.py *code from the book's website. This program uses the operating system's text-to-speech functionality in place of pyttsx3. You'll still need to update the Haar cascade path variable in this program, as you did at ❸ in Listing 9-1.*

Next, use the time module to pause the program for three seconds. This gives the intruder time to squarely face the gun's camera.

At this point, you'd make a video capture, except we're not using video. Instead, load the images in the *corridor_5* folder. Call the cv.imread() method with the IMREAD_GRAYSCALE flag ❸.

Use the image's shape attribute to get its height and width in pixels. This will come in handy later, when you post text on the images.

Face detection works only on grayscale images, but OpenCV will convert color images behind the scenes when applying the Haar cascades. I chose to use grayscale from the start as the results look creepier when the images display. If you want to see the images in color, just change the two previous lines as follows:

```
img_gray = cv.imread(image)
height, width = img_gray.shape[:2]
```

Next, show the image prior to face detection, keep it up for two seconds (input as milliseconds), and then destroy the window. This is for quality control to be sure all the images are being examined. You can comment out these steps later, after you're satisfied everything is working as planned.

Create an empty list to hold any faces found in the current image ❹. OpenCV treats images as NumPy arrays, so the items in this list are the corner-point coordinates (*x*, *y*, width, height) of a rectangle that frames the face, as shown in the following output snippet:

```
[array([[383, 169,  54,  54]], dtype=int32)]
```

Now it's time to detect faces using the Haar cascades. Do this for the face_cascade variable by calling the detectMultiscale() method. Pass the method the image and values for the scale factor and minimum number of neighbors. These can be used to tune the results in the event of false positives or failure to recognize faces.

For good results, the faces in an image should be the same size as the ones used to train the classifier. To ensure they are, the scaleFactor parameter rescales the original image to the correct size using a technique called a *scale pyramid* (Figure 9-6).

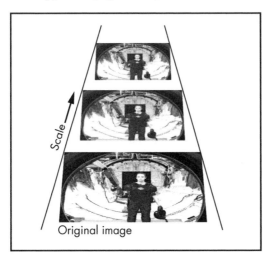

Figure 9-6: Example "scale pyramid"

The scale pyramid resizes the image downward a set number of times. For example, a scaleFactor of 1.2 means the image will be scaled down in increments of 20 percent. The sliding window will repeat its movement across this smaller image and check again for Haar features. This shrinking and sliding will continue until the scaled image reaches the size of the images used for training. This is 20×20 pixels for the Haar cascade classifier (you can confirm this by opening one of the *.xml* files). Windows smaller than this can't be detected, so the resizing ends at this point. Note that the scale pyramid will only *downscale* images, as upscaling can introduce artifacts in the resized image.

With each rescaling, the algorithm calculates lots of new Haar features, resulting in lots of false positives. To weed these out, use the minNeighbors parameter.

To see how this process works, look at Figure 9-7. The rectangles in this figure represent faces detected by the haarcascade_frontalface_alt2.xml classifier, with the scaleFactor parameter set to 1.05 and minNeighbors set to 0. The rectangles have different sizes depending on which scaled image— determined by the scaleFactor parameter—was in use when the face was detected. Although there are many false positives, the rectangles tend to cluster around the true face.

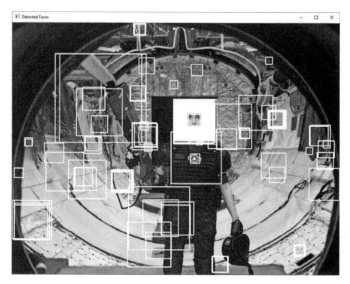

Figure 9-7: Detected face rectangles with `minNeighbors=0`

Increasing the value of the `minNeighbors` parameter will increase the quality of the detections but reduce their number. If you specify a value of 1, only rectangles with one or more closely neighboring rectangles are preserved, and all others are discarded (Figure 9-8).

Figure 9-8: Detected face rectangles with `minNeighbors=1`

Increasing the number of minimum neighbors to around five generally removes the false positives (Figure 9-9). This may be good enough for most applications, but dealing with terrifying interdimensional monstrosities demands more rigor.

Figure 9-9: Detected face rectangles with `minNeighbors=5`

To see why, check out Figure 9-10. Despite using a `minNeighbor` value of 5, the toe region of the mutant is incorrectly identified as a face. With a little imagination, you can see two dark eyes and a bright nose at the top of the rectangle, and a dark, straight mouth at the base. This could allow the mutant to pass unharmed, earning you a dishonorable discharge at best and an excruciatingly painful death at worst.

Figure 9-10: A mutant's right toe region incorrectly identified as a face

Fortunately, this problem can be easily remedied. The solution is to search for more than just faces.

Detecting Eyes and Disabling the Weapon

Still in the for loop through the corridor images, Listing 9-3 uses OpenCV's built-in eye cascade classifier to search for eyes in the list of detected face rectangles. Searching for eyes reduces false positives by adding a second verification step. And because mutants don't have eyes, if at least one eye is found, you can assume a human is present and disable the sentry gun's firing mechanism to let them pass.

sentry.py, part 3

```
print(f"Searching {image} for eyes.")
for rect in face_rect_list:
    for (x, y, w, h) in rect:
❶      rect_4_eyes = img_gray[y:y+h, x:x+w]
        eyes = eye_cascade.detectMultiScale(image=rect_4_eyes,
                                            scaleFactor=1.05,
                                            minNeighbors=2)
❷      for (xe, ye, we, he) in eyes:
            print("Eyes detected.")
            center = (int(xe + 0.5 * we), int(ye + 0.5 * he))
            radius = int((we + he) / 3)
            cv.circle(rect_4_eyes, center, radius, 255, 2)
            cv.rectangle(img_gray, (x, y), (x+w, y+h), (255, 255, 255), 2)
❸          discharge_weapon = False
            break
```

Listing 9-3: Detecting eyes in face rectangles and disabling the weapon

Print the name of the image being searched and start a loop through the rectangles in the face_rect_list. If a rectangle is present, start looping through the tuple of coordinates. Use these coordinates to make a subarray from the image, in which you'll search for eyes ❶.

Call the eye cascade classifier on the subarray. Because you're now searching a much smaller area, you can reduce the minNeighbors argument.

Like the cascade classifiers for faces, the eye cascade returns coordinates for a rectangle. Start a loop through these coordinates, naming them with an e on the end, which stands for "eye," to distinguish them from the face rectangle coordinates ❷.

Next, draw a circle around the first eye you find. This is just for your own visual confirmation; as far as the algorithm's concerned, the eye is already found. Calculate the center of the rectangle and then calculate a radius value that's slightly larger than an eye. Use OpenCV's circle() method to draw a white circle on the rect_4_eyes subarray.

Now, draw a rectangle around the face by calling OpenCV's rectangle() method and passing it the img_gray array. Show the image for two seconds and then destroy the window. Because the rect_4_eyes subarray is part of img_gray, the circle will show up even though you didn't explicitly pass the subarray to the im_show() method (Figure 9-11).

Figure 9-11: Face rectangle and eye circle

With a human identified, disable the weapon ❸ and break out of the for loop. You need to identify only one eye to confirm that you have a face, so it's time to move on to the next face rectangle.

Passing the Intruder or Discharging the Weapon

Still in the for loop through the corridor images, Listing 9-4 determines what happens if the weapon is disabled or if it's allowed to fire. In the disabled case, it shows the image with the detected face and plays the "all clear" tone. Otherwise, it shows the image and plays the gunfire audio file.

sentry.py, part 4

```
if discharge_weapon == False:
    playsound(tone_path, block=False)
    cv.imshow('Detected Faces', img_gray)
    cv.waitKey(2000)
    cv.destroyWindow('Detected Faces')
    time.sleep(5)

else:
    print(f"No face in {image}. Discharging weapon!")
    cv.putText(img_gray, 'FIRE!', (int(width / 2) - 20, int(height / 2)),
                        cv.FONT_HERSHEY_PLAIN, 3, 255, 3)
    playsound(gunfire_path, block=False)
    cv.imshow('Mutant', img_gray)
    cv.waitKey(2000)
    cv.destroyWindow('Mutant')
    time.sleep(3)

engine.stop()
```

Listing 9-4: Determining the course of action if the gun is disabled or enabled

Use a conditional to check whether the weapon is disabled. You set the discharge_weapon variable to True when you chose the current image from the *corridor_5* folder (see Listing 9-2). If the previous listing found an eye in a face rectangle, it changed the state to False.

If the weapon is disabled, show the positive detection image (such as in Figure 9-11) and play the tone. First, call playsound, pass it the tone_path string, and set the block argument to False. By setting block to False, you allow playsound to run at the same time as OpenCV displays the image. If you set block=True, you won't see the image until *after* the tone audio has completed. Show the image for two seconds and then destroy it and pause the program for five seconds using time.sleep().

If discharge_weapon is still True, print a message to the shell that the gun is firing. Use OpenCV's putText() method to announce this in the center of the image and then show the image (see Figure 9-12).

Figure 9-12: Example mutant window

Now play the gunfire audio. Use playsound, passing it the gunfire_path string and setting the block argument to False. Note that you have the option of removing the root_dir and gunfire_path lines of code in Listing 9-1 if you provide the full path when you call playsound. For example, I would use the following on my Windows machine:

```
playsound('C:/Python372/book/mutants/gunfire.wav', block=False)
```

Show the window for two seconds and then destroy it. Sleep the program for three seconds to pause between showing the mutant and displaying the next image in the *corridor_5* folder. When the loop completes, stop the pyttsx3 engine.

Results

Your *sentry.py* program repaired the damage to the sentry gun and allowed it to function without the need for transponders. It's biased to preserve human life, however, which could lead to disastrous consequences: if a mutant enters the corridor at around the same time as a human, the mutant could slip by the defenses (Figure 9-13).

Figure 9-13: A worst-case scenario. Say "Cheese!"

Mutants might also trigger the firing mechanism with humans in the corridor, assuming the humans look away from the camera at the wrong moment (Figure 9-14).

Figure 9-14: You had one job!

I've seen enough sci-fi and horror movies to know that in a real scenario, I'd program the gun to shoot anything that moved. Fortunately, that's a moral dilemma I'll never have to *face*!

Detecting Faces from a Video Stream

You can also detect faces in real time using video cameras. This is easy to do, so we won't make it a dedicated project. Enter the code in Listing 9-5 or use the digital version, *video_face_detect.py*, in the *Chapter_9* folder download-able from the book's website. You'll need to use your computer's camera or an external camera that works through your computer.

---

video_face_detect.py
```
import cv2 as cv

path = "C:/Python372/Lib/site-packages/cv2/data/"
face_cascade = cv.CascadeClassifier(path + 'haarcascade_frontalface_alt.xml')

❶ cap = cv.VideoCapture(0)

while True:
    _, frame = cap.read()
    face_rects = face_cascade.detectMultiScale(frame, scaleFactor=1.2,
                                               minNeighbors=3)

    for (x, y, w, h) in face_rects:
        cv.rectangle(frame, (x, y), (x+w, y+h), (0, 255, 0), 2)

    cv.imshow('frame', frame)
❷ if cv.waitKey(1) & 0xFF == ord('q'):
        break

cap.release()
cv.destroyAllWindows()
```

Listing 9-5: Detecting faces in a video stream

After importing OpenCV, set up your path to the Haar cascade classifi-ers as you did at ❸ in Listing 9-1. I use the *haarcascade_frontalface_alt.xml* file here as it has higher precision (fewer false positives) than the *haarcascade _frontalface_default.xml* file you used in the previous project. Next, instantiate a VideoCapture class object, called cap for "capture." Pass the constructor the index of the video device you want to use ❶. If you have only one camera, such as your laptop's built-in camera, then the index of this device should be 0.

To keep the camera and face detection process running, use a while loop. Within the loop, you'll capture each video frame and analyze it for faces, just as you did with the static images in the previous project. The face detection algorithm is fast enough to keep up with the continuous stream, despite all the work it must do!

To load the frames, call the cap object's read() method. It returns a tuple consisting of a Boolean return code and a NumPy ndarray object representing the current frame. The return code is used to check whether you've run out of frames when reading from a file. Since we're not reading from a file here, assign it to an underscore to indicate an insignificant variable.

Next, reuse the code from the previous project that finds face rectangles and draws the rectangles on the frame. Display the frame with the OpenCV imshow() method. The program should draw a rectangle on this frame if it detects a face.

To end the loop, you'll press the Q key, for quit ❷. Start by calling OpenCV's waitKey() method and passing it a short, one-millisecond timespan. This method will pause the program as it waits for a key press, but we don't want to interrupt the video stream for too long.

Python's built-in ord() function accepts a string as an argument and returns the Unicode code point representation of the passed argument, in this case a lowercase *q*. You can see a mapping of characters to numbers here: *http://www.asciitable.com/*. To make this lookup compatible with all operating systems, you must include the bitwise AND operator, &, with the hexadecimal number FF (0xFF), which has an integer value of 255. Using & 0xFF ensures only the last 8 bits of the variable are read.

When the loop ends, call the cap object's release() method. This frees up the camera for other applications. Complete the program by destroying the display window.

You can add more cascades to the face detection to increase its accuracy, as you did in the previous project. If this slows detection too much, try scaling down the video image. Right after the call to cap.read(), add the following snippet:

```
frame = cv.resize(frame, None, fx=0.5, fy=0.5,
                  interpolation=cv.INTER_AREA)
```

The fx and fy arguments are scaling factors for the screen's *x* and *y* dimensions. Using 0.5 will halve the default size of the window.

The program should have no trouble tracking your face unless you do something crazy, like tilt your head slightly to the side. That's all it takes to break detection and make the rectangle disappear (Figure 9-15).

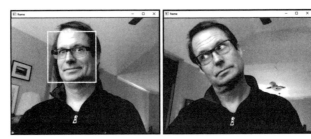

Figure 9-15: Face detection using video frames

Haar cascade classifiers are designed to recognize upright faces, both frontal and profile views, and they do a great job. They can even handle eyeglasses and beards. But tilt your head, and they can quickly fail.

An inefficient but simple way to manage tilted heads is to use a loop that rotates the images slightly before passing them on for face detection.

The Haar cascade classifiers can handle a bit of tilt (Figure 9-16), so you could rotate the image by 5 degrees or so with each pass and have a good chance of getting a positive result.

Figure 9-16: Rotating the image facilitates face detection.

The Haar feature approach to face detection is popular because it's fast enough to run in real time with limited computational resources. As you probably suspect, however, more accurate, sophisticated, and resource-intensive techniques are available.

For example, OpenCV ships with an accurate and robust face detector based on the Caffe deep learning framework. To learn more about this detector, see the tutorial "Face Detection with OpenCV and Deep Learning" at *https://www.pyimagesearch.com/*.

Another option is to use OpenCV's LBP cascade classifier for face detection. This technique divides a face into blocks and then extracts local binary pattern histograms (LBPHs) from them. Such histograms have proved effective at detecting *unconstrained* faces in images—that is, faces that aren't well aligned and with similar poses. We'll look at LBPH in the next chapter, where we'll focus on *recognizing* faces rather than simply detecting them.

Summary

In this chapter, you got to work with OpenCV's Haar cascade classifier for detecting human faces; playsound, for playing audio files; and pyttsx3, for text-to-speech audio. Thanks to these useful libraries, you were able to quickly write a face detection program that also issued audio warnings and instructions.

Further Reading

"Rapid Object Detection Using a Boosted Cascade of Simple Features" (Conference on Computer Vision and Pattern Recognition, 2001), by Paul Viola and Michael Jones, is the first object detection framework to provide practical, real-time object detection rates. It forms the basis for the face detection process used in this chapter.

Adrian Rosebrock's *https://www.pyimagesearch.com/* website is an excellent source for building image search engines and finding loads of interesting computer vision projects, such as programs that detect fire and smoke, find targets in drone video streams, distinguish living faces from printed faces, automatically recognize license plates, and do much, much more.

Practice Project: Blurring Faces

Have you ever seen a documentary or news report where a person's face has been blurred to preserve their anonymity, like in Figure 9-17? Well, this cool effect is easy to do with OpenCV. You just need to extract the face rectangle from a frame, blur it, and then write it back over the frame image, along with an (optional) rectangle outlining the face.

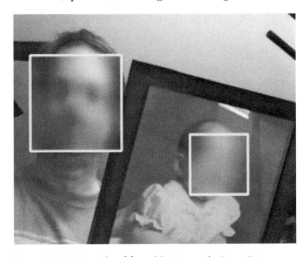

Figure 9-17: Example of face blurring with OpenCV

Blurring averages pixels within a local matrix called a *kernel*. Think of the kernel as a box you place on an image. All the pixels in this box are averaged to a single value. The larger the box, the more pixels are averaged, and thus the smoother the image appears. Thus, you can think of blurring as a low-pass filter that blocks high-frequency content, such as sharp edges.

Blurring is the only step in this process you haven't done before. To blur an image, use the OpenCV blur() method and pass it an image and a tuple of the kernel size in pixels.

```
blurred_image = cv.blur(image, (20, 20))
```

In this example, you replace the value of a given pixel in `image` with the average of all the pixels in a 20×20 square centered on that pixel. This operation repeats for every pixel in `image`.

You can find a solution, *practice_blur.py*, in the appendix and in the *Chapter_9* folder downloadable from the book's website.

Challenge Project: Detecting Cat Faces

It turns out there are three animal life forms on planet LV-666: humans, mutants, and cats. The base's mascot, Mr. Kitty, has free rein of the place and is prone to wander through Corridor 5.

Edit and calibrate *sentry.py* so that Mr. Kitty can freely pass. This will be a challenge, as cats aren't known for obeying verbal orders. To get him to at least look at the camera, you might add a "Here kitty, kitty" or "Puss, puss, puss" to the `pyttsx3` verbal commands. Or better, add the sound of a can of tuna being opened using `playsound`!

You can find Haar classifiers for cat faces in the same OpenCV folder as the classifiers you used in Project 13, and an empty corridor image, *empty_corridor.png*, in the book's downloadable *Chapter_9* folder. Select a few cat images from the internet, or your personal collection, and paste them in different places in the empty corridor. Use the humans in the other images to gauge the proper scale for the cat.

10

RESTRICTING ACCESS WITH FACE RECOGNITION

In the previous chapter, you were a technician in the Coalition Marines, a branch of the Space Force. In this chapter, you're that same technician, only your job just got harder. Your role is now to *recognize* faces, rather than just detect them. Your commander, Captain Demming, has discovered the lab containing the mutant-producing interdimensional portal, and he wants access to it restricted to just himself.

As in the previous chapter, you'll need to act quickly, so you'll rely on Python and OpenCV for speed and efficiency. Specifically, you'll use OpenCV's local binary pattern histogram (LBPH) algorithm, one of the oldest and easiest to use face recognition algorithms, to help lock down the lab. If you haven't installed and used OpenCV before, check out "Installing the Python Libraries" on page 6.

Recognizing Faces with Local Binary Pattern Histograms

The LBPH algorithm relies on feature vectors to recognize faces. Remember from Chapter 5 that a feature vector is basically a list of numbers in a specific order. In the case of LBPH, the numbers represent some qualities of a face. For instance, suppose you could discriminate between faces with just a few measurements, such as the separation of the eyes, the width of the mouth, the length of the nose, and the width of the face. These four measurements, in the order listed and expressed in centimeters, could compose the following feature vector: (5.1, 7.4, 5.3, 11.8). Reducing faces in a database to these vectors enables rapid searches, and it allows us to express the difference between them as the numerical difference, or *distance*, between two vectors.

Recognizing faces computationally requires more than four features, of course, and the many available algorithms work on different features. Among these algorithms are Eigenfaces, LBPH, Fisherfaces, scale-invariant feature transform (SIFT), speeded-up robust features (SURF), and various neural network approaches. When the face images are acquired under controlled conditions, these algorithms can have a high accuracy rate, about as high as that of humans.

Controlled conditions for images of faces might involve a frontal view of each face with a normal, relaxed expression and, to be usable by all algorithms, consistent lighting conditions and resolutions. The face should be unobscured by facial hair and glasses, assuming the algorithm was taught to recognize the face under those conditions.

The Face Recognition Flowchart

Before getting into the details of the LBPH algorithm, let's look at how face recognition works in general. The process consists of three main steps: capturing, training, and predicting.

In the capture phase, you gather the images that you'll use to train the face recognizer (Figure 10-1). For each face you want to recognize, you should take a dozen or more images with multiple expressions.

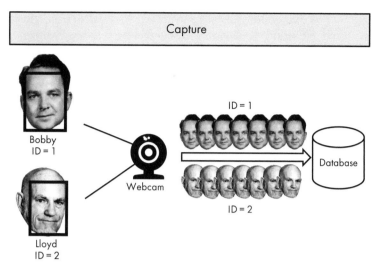

Figure 10-1: Capturing facial images to train the face recognizer

The next step in the capture process is to detect the face in the image, draw a rectangle around it, crop the image to the rectangle, resize the cropped images to the same dimensions (depending on the algorithm), and convert them to grayscale. The algorithms typically keep track of faces using integers, so each subject will need a unique ID number. Once processed, the faces are stored in a single folder, which we'll call the database.

The next step is to train the face recognizer (Figure 10-2). The algorithm—in our case, LBPH—analyzes each of the training images and then writes the results to a YAML (*.yml*) file, a human-readable data-serialization language used for data storage. YAML originally meant "Yet Another Markup Language" but now stands for "YAML Ain't Markup Language" to stress that it's more than just a document markup tool.

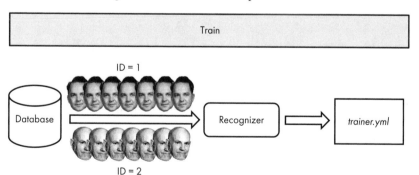

Figure 10-2: Training the face recognizer and writing the results to a file

With the face recognizer trained, the final step is to load a new, untrained face and predict its identity (Figure 10-3). These unknown faces are prepped in the same manner as the training images—that is, cropped, resized, and converted to grayscale. The recognizer then analyzes them, compares the results to the faces in the YAML file, and predicts which face matches best.

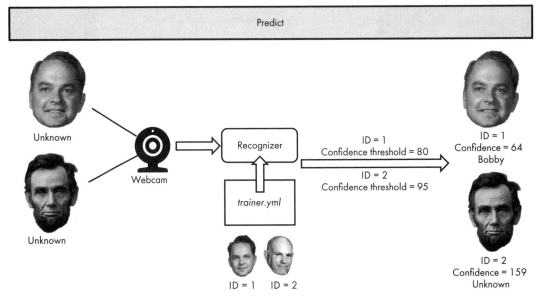

Figure 10-3: Predicting unknown faces using the trained recognizer

Note that the recognizer will make a prediction about the identity of every face. If there's only one trained face in the YAML file, the recognizer will assign every face the trained face's ID number. It will also output a *confidence* factor, which is really a measurement of the distance between the new face and the trained face. The larger the number, the worse the match. We'll talk about this more in a moment, but for now, know that you'll use a threshold value to decide whether the predicted face is correct. If the confidence exceeds the accepted threshold value, the program will discard the match and classify the face as "unknown" (see Figure 10-3).

Extracting Local Binary Pattern Histograms

The OpenCV face recognizer you'll use is based on local binary patterns. These texture descriptors were first used around 1994 to describe and classify surface textures, differentiating concrete from carpeting, for example. Faces are also composed of textures, so the technique works for face recognition.

Before you can extract histograms, you first need to generate the binary patterns. An LBP algorithm computes a local representation of texture by comparing each pixel with its surrounding neighbors. The first computational step is to slide a small window across the face image and capture the pixel information. Figure 10-4 shows an example window.

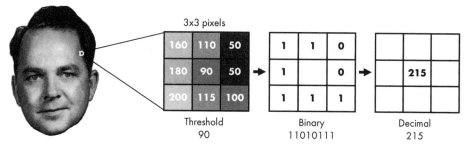

Figure 10-4: Example 3x3 pixel sliding window used to capture local binary patterns

The next step is to convert the pixels into a binary number, using the central value (in this case 90) as a threshold. You do this by comparing the eight neighboring values to the threshold. If a neighboring value is equal to or higher than the threshold, assign it 1; if it's lower than the threshold, assign it 0. Next, ignoring the central value, concatenate the binary values line by line (some methods use a clockwise rotation) to form a new binary value (11010111). Finish by converting this binary number into a decimal number (215) and storing it at the central pixel location.

Continue sliding the window until all the pixels have been converted to LBP values. In addition to using a square window to capture neighboring pixels, the algorithm can use a radius, a process called *circular LBP*.

Now it's time to extract histograms from the LBP image produced in the previous step. To do this, you use a grid to divide the LBP image into rectangular regions (Figure 10-5). Within each region, you construct a histogram of the LBP values (labeled "Local Region Histogram" in Figure 10-5).

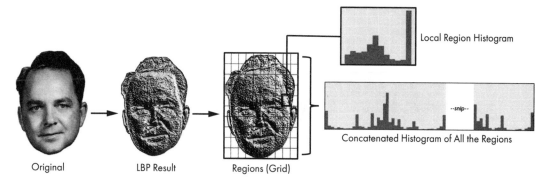

Figure 10-5: Extracting the LBP histograms

After constructing the local region histograms, you follow a predetermined order to normalize and concatenate them into one long histogram (shown truncated in Figure 10-5). Because you're using a grayscale image with intensity values between 0 and 255, there are 256 positions in each histogram. If you're using a 10×10 grid, as in Figure 10-5, then there are 10×10×256 = 25,600 positions in the final histogram. The assumption is that this composite histogram includes diagnostic features needed to recognize a face. They are thus *representations* of a face image, and face recognition consists of comparing these representations, rather than the images themselves.

To predict the identity of a new, unknown face, you extract its concatenated histogram and compare it to the existing histograms in the trained database. The comparison is a measure of the distance between histograms. This calculation may use various methods, including Euclidian distance, absolute distance, chi-square, and so on. The algorithm returns the ID number of the trained image with the closest histogram match, along with the confidence measurement. You can then apply a threshold to the confidence value, as in Figure 10-3. If the confidence for the new image is below the threshold value, assume you have a positive match.

Because OpenCV encapsulates all these steps, the LBPH algorithm is easy to implement. It also produces great results in a controlled environment and is unaffected by changes in lighting conditions (Figure 10-6).

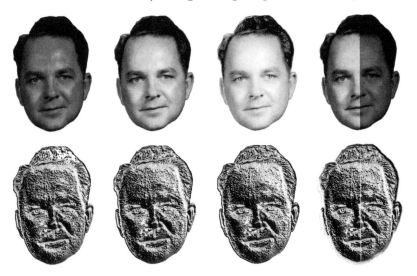

Figure 10-6: LBPs are robust against changes in illumination

The LBPH algorithm handles changes to lighting conditions well because it relies on comparisons among pixel intensities. Even if illumination is much brighter in one image than another, the relative reflectivity of the face remains the same, and LBPH can capture it.

Project #14: Restricting Access to the Alien Artifact

Your squad has fought its way to the lab containing the portal-producing alien artifact. Captain Demming orders it locked down immediately, with access restricted to just him. Another technician will override the current system with a military laptop. Captain Demming will gain access through this laptop using two levels of security: a typed password and face verification. Aware of your skills with OpenCV, he's ordered *you* to handle the facial verification part.

THE OBJECTIVE

Write a Python program that recognizes Captain Demming's face.

The Strategy

You're pressed for time and working under adverse conditions, so you want to use a fast and easy tool with a good performance record, like OpenCV's LBPH face recognizer. You're aware that LBPH works best under controlled conditions, so you'll use the same laptop webcam to capture both the training images and the face of anyone trying to access the lab.

In addition to pictures of Demming's face, you'll want to capture some faces that don't belong to Captain Demming. You'll use these faces to ensure that all the positive matches really belong to the captain. Don't worry about setting up the password, isolating the program from the user, or hacking into the current system; the other technician will handle these tasks while you go out and blast some mutants.

Supporting Modules and Files

You'll use both OpenCV and NumPy to do most of the work in this project. If you don't already have them installed, see "Installing the Python Libraries" on page 6. You'll also need playsound, for playing sounds, and pyttsx3, for text-to-speech functionality. You can find out more about these modules, including installation instructions, in "The Code" on page 207.

The code and supporting files are in the *Chapter_10* folder from the book's website, *https://nostarch.com/real-world-python/*. Keep the folder structure and filenames the same after downloading them (Figure 10-7). Note that the *tester* and *trainer* folders are created later and will not be included in the download.

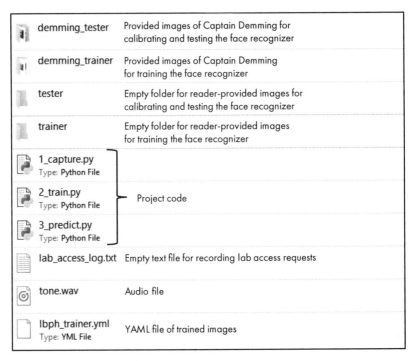

![]	demming_tester	Provided images of Captain Demming for calibrating and testing the face recognizer
![]	demming_trainer	Provided images of Captain Demming for training the face recognizer
![]	tester	Empty folder for reader-provided images for calibrating and testing the face recognizer
![]	trainer	Empty folder for reader-provided images for training the face recognizer
![]	1_capture.py Type: Python File	
![]	2_train.py Type: Python File	Project code
![]	3_predict.py Type: Python File	
![]	lab_access_log.txt	Empty text file for recording lab access requests
![]	tone.wav	Audio file
![]	lbph_trainer.yml Type: YML File	YAML file of trained images

Figure 10-7: File structure for Project 14

The *demming_trainer* and *demming_tester* folders contain images of Captain Demming and others that you can use for this project. The code currently references these folders.

If you want to supply your own images—for example, to use your own face to represent Captain Demming's—then you'll use the folders named *trainer* and *tester*. The code that follows will create the *trainer* folder for you. You'll need to manually create the *tester* folder and add some images of yourself, as described later. Of course, you'll need to edit the code so that it points to these new folders.

The Video Capture Code

The first step (performed by the *1_capture.py* code) is to capture the facial images that you'll need for training the recognizer. You can skip this step if you plan to use the images provided in the *demming_trainer* folder.

To use your own face for Captain Demming, use your computer's camera to record about a dozen face shots with various expressions and no glasses. If you don't have a webcam, you can skip this step, take selfies with your phone, and save them to a folder named *trainer*, as shown in Figure 10-7.

Importing Modules, and Setting Up Audio, a Webcam, Instructions, and File Paths

Listing 10-1 imports modules, initializes and sets up the audio engine and the Haar cascade classifier, initializes the camera, and provides user instructions. You need the Haar cascades because you must detect a face before you can recognize it. For a refresher on Haar cascades and face detection, see "Detecting Faces in Photographs" on page 204.

1_capture.py,
part 1

```
import os
import pyttsx3
import cv2 as cv
from playsound import playsound

engine = pyttsx3.init()
❶ engine.setProperty('rate', 145)
engine.setProperty('volume', 1.0)

root_dir = os.path.abspath('.')
tone_path = os.path.join(root_dir, 'tone.wav')

❷ path = "C:/Python372/Lib/site-packages/cv2/data/"
face_detector = cv.CascadeClassifier(path +
                             'haarcascade_frontalface_default.xml')

cap = cv.VideoCapture(0)
if not cap.isOpened():
    print("Could not open video device.")
❸ cap.set(3, 640)  # Frame width.
cap.set(4, 480)  # Frame height.

engine.say("Enter your information when prompted on screen. \
           Then remove glasses and look directly at webcam. \
           Make multiple faces including normal, happy, sad, sleepy. \
           Continue until you hear the tone.")
engine.runAndWait()

❹ name = input("\nEnter last name: ")
user_id = input("Enter assigned ID Number: ")
print("\nCapturing face. Look at the camera now!")
```

Listing 10-1: Importing modules and setting up audio and detector files, a webcam, and instructions

The imports are the same as those used to detect faces in the previous chapter. You'll use the operating system (via the os module) to manipulate file paths, pyttsx3 to play text-to-speech audio instructions, cv to work with images and run the face detector and recognizer, and playsound to play a tone that lets users know when the program has finished capturing their image.

Next, set up the text-to-speech engine. You'll use this to tell the user how to run the program. The default voice is dependent on your particular operating system. The engine's rate parameter is currently optimized for the American "David" voice on Windows ❶. You may want to edit the argument if you find the speech to be too fast or too slow. If you want to change the voice, see the instructions accompanying Listing 9-1 on page 209.

You'll use a tone to alert the user that the video capture process has ended. Set up the path to the *tone.wav* audio file as you did in Chapter 9.

Now, provide the path to the Haar cascade file ❷ and assign the classifier to a variable named face_detector. The path shown here is for my Windows machine; your path may be different. On macOS, for example, you can find the files under *opencv/data/haarcascades*. You can also find them online at *https://github.com/opencv/opencv/tree/master/data/haarcascades/*.

In Chapter 9, you learned how to capture your face using your computer's webcam. You'll use similar code in this program, starting with a call to cv.VideoCapture(0). The 0 argument refers to the active camera. If you have multiple cameras, you may need to use another number, such as 1, which you can determine through trial and error. Use a conditional to check that the camera opened, and if it did, set the frame width and height, respectively ❸. The first argument in both methods refers to the position of the width or height parameter in the list of arguments.

For security reasons, you'll be present to supervise the video capture phase of the process. Nevertheless, use the pyttsx3 engine to explain the procedure to the user (this way you don't have to remember it). To control the acquisition conditions to ensure accurate recognition later, the user will need to remove any glasses or face coverings and adopt multiple expressions. Among these should be the expression they plan to use each time they access the lab.

Finally, they'll need to follow some printed instructions on the screen. First, they'll enter their last name ❹. You don't need to worry about duplicates right now, as Captain Demming will be the only user. Plus, you'll assign the user a unique ID number. OpenCV will use this variable, user_id, to keep track of all the faces during training and prediction. Later, you'll create a dictionary so you can keep track of which user ID belongs to which person, assuming more people are granted access in the future.

As soon as the user enters their ID number and presses ENTER, the camera will turn on and begin capturing images, so let them know this with another call to print(). Remember from the previous chapter that the Haar cascade face detector is sensitive to head orientation. For it to function properly, the user must look right at the webcam and keep their head as straight as possible.

Capturing the Training Images

Listing 10-2 uses the webcam and a while loop to capture a specified number of face images. The code saves the images to a folder and sounds a tone when the operation is complete.

1_capture.py, part 2

```
if not os.path.isdir('trainer'):
    os.mkdir('trainer')
os.chdir('trainer')

frame_count = 0

while True:
    # Capture frame-by-frame for total of 30 frames.
    _, frame = cap.read()
    gray = cv.cvtColor(frame, cv.COLOR_BGR2GRAY)
```

```
❶ face_rects = face_detector.detectMultiScale(gray, scaleFactor=1.2,
                                                 minNeighbors=5)
   for (x, y, w, h) in face_rects:
       frame_count += 1
       cv.imwrite(str(name) + '.' + str(user_id) + '.'
                   + str(frame_count) + '.jpg', gray[y:y+h, x:x+w])
       cv.rectangle(frame, (x, y), (x + w, y + h), (0, 255, 0), 2)
       cv.imshow('image', frame)
       cv.waitKey(400)
❷ if frame_count >= 30:
       break

print("\nImage collection complete. Exiting...")
playsound(tone_path, block=False)
cap.release()
cv.destroyAllWindows()
```

Listing 10-2: Capturing video images using a loop

Start by checking for a directory named *trainer*. If it doesn't exist, use the operating system module's mkdir() method to make the directory. Then change the current working directory to this *trainer* folder.

Now, initialize a frame_count variable to 0. The code will capture and save a video frame only if it detects a face. To know when to end the program, you'll need to keep count of the number of captured frames.

Next, start a while loop set to True. Then call the cap object's read() method. As noted in the previous chapter, this method returns a tuple consisting of a Boolean return code and a numpy ndarray object representing the current frame. The return code is typically used to check whether you've run out of frames when reading from a file. Since we're not reading from a file here, assign it to an underscore to indicate an unused variable.

Both face detection and face recognition work on grayscale images, so convert the frame to grayscale and name the resulting array gray. Then, call the detectMultiscale() method to detect faces in the image ❶. You can find details of how this method works in the discussion of Listing 9-2 on page 212. Because you're controlling conditions by having the user look into a laptop's webcam, you can rest assured that the algorithm will work well, though you should certainly check the results.

The previous method should output the coordinates for a rectangle around the face. Start a for loop through each set of coordinates and immediately advance the frame_count variable by 1.

Use OpenCV's imwrite() method to save the image to the *trainer* folder. The folders use the following naming logic: *name.user_id.frame_count.jpg* (such as *demming.1.9.jpg*). Save only the portion of the image within the face rectangle. This will help ensure you aren't training the algorithm to recognize background features.

The next two lines draw a face rectangle on the original frame and show it. This is so the user—Captain Demming—can check that his head is upright and his expressions are suitable. The waitKey() method delays the capture process enough for the user to cycle through multiple expressions.

Even if Captain Demming will always adopt a relaxed, neutral expression when having his identity verified, training the software on a range of expressions will lead to more robust results. Along these lines, it's also helpful if the user tilts their head *slightly* from side to side during the capture phase.

Next, check whether the target frame count has been reached, and if it has, break out of the loop ❷. Note that, if no one is looking at the camera, the loop will run forever. It counts frames only if the cascade classifier detects a face and returns a face rectangle.

Let the user know that the camera has turned off by printing a message and sounding the tone. Then end the program by releasing the camera and destroying all the image windows.

At this point, the *trainer* folder should contain 30 images of the user's closely cropped face. In the next section, you'll use these images—or the set provided in the *demming_trainer* folder—to train OpenCV's face recognizer.

The Face Trainer Code

The next step is to use OpenCV to create an LBPH-based face recognizer, train it with the training images, and save the results as a reusable file. If you're using your own face to represent Captain Demming's, you'll point the program to the *trainer* folder. Otherwise, you'll need to use the *demming _trainer* folder, which, along with the *2_train.py* file containing the code, is in the downloadable *Chapter_10* folder.

Listing 10-3 sets up paths to the Haar cascades used for face detection and the training images captured by the previous program. OpenCV keeps track of faces using label integers, rather than name strings, and the listing also initializes lists to hold the labels and their related images. It then loops through the training images, loads them, extracts a user ID number from the filename, and detects the faces. Finally, it trains the recognizer and saves the results to a file.

2_train.py

```
import os
import numpy as np
import cv2 as cv

cascade_path = "C:/Python372/Lib/site-packages/cv2/data/"
face_detector = cv.CascadeClassifier(cascade_path +
                                     'haarcascade_frontalface_default.xml')

❶ train_path = './demming_trainer'  # Use for provided Demming face.
  #train_path = './trainer'  # Uncomment to use your face.
  image_paths = [os.path.join(train_path, f) for f in os.listdir(train_path)]
  images, labels = [], []

  for image in image_paths:
      train_image = cv.imread(image, cv.IMREAD_GRAYSCALE)
❷     label = int(os.path.split(image)[-1].split('.')[1])
      name = os.path.split(image)[-1].split('.')[0]
      frame_num = os.path.split(image)[-1].split('.')[2]
❸     faces = face_detector.detectMultiScale(train_image)
      for (x, y, w, h) in faces:
```

```
        images.append(train_image[y:y + h, x:x + w])
        labels.append(label)
        print(f"Preparing training images for {name}.{label}.{frame_num}")
        cv.imshow("Training Image", train_image[y:y + h, x:x + w])
        cv.waitKey(50)

cv.destroyAllWindows()

❹ recognizer = cv.face.LBPHFaceRecognizer_create()
recognizer.train(images, np.array(labels))
recognizer.write('lbph_trainer.yml')
print("Training complete. Exiting...")
```

Listing 10-3: Training and saving the LBPH face recognizer

You've seen the imports and the face detector code before. Although you've already cropped the training images to face rectangles in *1_capture.py*, it doesn't hurt to repeat this procedure. Since *2_train.py* is a stand-alone program, it's best not to take anything for granted.

Next, you must choose which set of training images to use: the ones you captured yourself in the *trainer* folder or the set provided in the *demming _trainer* folder ❶. Comment out or delete the line for the one you don't use. Remember, because you're not providing a full path to the folder, you'll need to launch your program from the folder containing it, which should be one level above the *trainer* and *demming_trainer* folders.

Create a list named image_paths using list comprehension. This will hold the directory path and filename for each image in the training folder. Then create empty lists for the images and their labels.

Start a for loop through the image paths. Read the image in grayscale; then extract its numeric label from the filename and convert it to an integer ❷. Remember that the label corresponds to the user ID input through *1_capture.py* right before it captured the video frames.

Let's take a moment to unpack what's happening in this extraction and conversion process. The os.path.split() method takes a directory path and returns a tuple of the directory path and the filename, as shown in the following snippet:

```
>>> import os
>>> path = 'C:\demming_trainer\demming.1.5.jpg'
>>> os.path.split(path)
('C:\\demming_trainer', 'demming.1.5.jpg')
```

You can then select the last item in the tuple, using an index of -1, and split it on the dot. This yields a list with four items (the user's name, user ID, frame number, and file extension).

```
>>> os.path.split(path)[-1].split('.')
['demming', '1', '5', 'jpg']
```

To extract the label value, you choose the second item in the list using index 1.

```
>>> os.path.split(path)[-1].split('.')[1]
'1'
```

Repeat this process to extract the name and frame_num for each image. These are all strings at this point, which is why you need to turn the user ID into an integer for use as a label.

Now, call the face detector on each training image ❸. This will return a numpy.ndarray, which you'll call faces. Start looping through the array, which contains the coordinates of the detected face rectangles. Append the part of the image in the rectangle to the images list you made earlier. Also append the image's user ID to the labels list.

Let the user know what's going on by printing a message in the shell. Then, as a check, show each training image for 50 milliseconds. If you've ever seen Peter Gabriel's popular 1986 music video for "Sledgehammer," you'll appreciate this display.

It's time to train the face recognizer. Just as you do when using OpenCV's face detector, you start by instantiating a recognizer object ❹. Next, you call the train() method and pass it the images list and the labels list, which you turn into a NumPy array on the fly.

You don't want to train the recognizer every time someone verifies their face, so write the results of the training process to a file called *lbph_trainer.yml*. Then let the user know the program has ended.

The Face Predictor Code

It's time to start recognizing faces, a process we'll call *predicting*, because it all comes down to probability. The program in *3_predict.py* will first calculate the concatenated LBP histogram for each face. It will then find the distance between this histogram and all the histograms in the training set. Next, it will assign the new face the label and name of the trained face that's closest to it, but only if the distance falls below a threshold value that you specify.

Importing Modules and Preparing the Face Recognizer

Listing 10-4 imports modules, prepares a dictionary to hold user ID numbers and names, sets up the face detector and recognizer, and establishes the path to the test data. The test data includes images of Captain Demming, along with several other faces. An image of Captain Demming from the training folder is included to test the results. If everything is working as it should, the algorithm should positively identify this image with a low distance measurement.

3_predict.py, part 1
```
import os
from datetime import datetime
import cv2 as cv

names = {1: "Demming"}
```

```
      cascade_path = "C:/Python372/Lib/site-packages/cv2/data/"
      face_detector = cv.CascadeClassifier(cascade_path +
                                  'haarcascade_frontalface_default.xml')

❶ recognizer = cv.face.LBPHFaceRecognizer_create()
   recognizer.read('lbph_trainer.yml')

   #test_path = './tester'
❷ test_path = './demming_tester'
   image_paths = [os.path.join(test_path, f) for f in os.listdir(test_path)]
```

Listing 10-4: Importing modules and preparing for face detection and recognition

After some familiar imports, create a dictionary to link user ID numbers to usernames. Although there's only one entry currently, this name dictionary makes it easy to add more entries in the future. If you're using your own face, feel free to change the last name, but leave the ID number set to 1.

Next, repeat the code that sets up the face_detector object. You'll need to input your own cascade_path (see Listing 10-1 on page 233).

Create a recognizer object as you did in the *2_train.py* code ❶. Then use the read() method to load the *.yml* file that contains the training information.

You'll want to test the recognizer using face images in a folder. If you're using the Demming images provided, set up a path to the *demming_tester* folder ❷. Otherwise, use the *tester* folder you created earlier. You can add your own images to this blank folder. If you're using your face to represent Captain Demming's, you shouldn't reuse the training images here, although you might consider using one as a control. Instead, use the *1_capture.py* program to produce some new images. If you wear glasses, include some images of you with and without them. You'll want to include some strangers from the *demming_tester* folder, as well.

Recognizing Faces and Updating an Access Log

Listing 10-5 loops through the images in the test folder, detects any faces present, compares the face histogram to those in the training file, names the face, assigns a confidence value, and then logs the name and access time in a persistent text file. As part of this process, the program would theoretically unlock the lab if the ID is positive, but since we don't have a lab, we'll skip that part.

3_predict.py, part 2
```
for image in image_paths:
    predict_image = cv.imread(image, cv.IMREAD_GRAYSCALE)
    faces = face_detector.detectMultiScale(predict_image,
                                   scaleFactor=1.05,
                                   minNeighbors=5)
    for (x, y, w, h) in faces:
        print(f"\nAccess requested at {datetime.now()}.")
❶      face = cv.resize(predict_image[y:y + h, x:x + w], (100, 100))
        predicted_id, dist = recognizer.predict(face)
❷      if predicted_id == 1 and dist <= 95:
            name = names[predicted_id]
            print("{} identified as {} with distance={}"
```

```
                .format(image, name, round(dist, 1)))
    ❸ print(f"Access granted to {name} at {datetime.now()}.",
            file=open('lab_access_log.txt', 'a'))
        else:
            name = 'unknown'
            print(f"{image} is {name}.")

        cv.rectangle(predict_image, (x, y), (x + w, y + h), 255, 2)
        cv.putText(predict_image, name, (x + 1, y + h - 5),
                   cv.FONT_HERSHEY_SIMPLEX, 0.5, 255, 1)
        cv.imshow('ID', predict_image)
        cv.waitKey(2000)
        cv.destroyAllWindows()
```

Listing 10-5: Running face recognition and updating the access log file

Start by looping through the images in the test folder. This will be either the *demming_tester* folder or the *tester* folder. Read each image in as grayscale and assign the resulting array to a variable named predict_image. Then run the face detector on it.

Now loop through the face rectangles, as you've done before. Print a message about access being requested; then use OpenCV to resize the face subarray to 100×100 pixels ❶. This is close to the dimensions of the training images in the *demming_trainer* folder. Synchronizing the size of the images isn't strictly necessary but helps to improve results in my experience. If you're using your own images to represent Captain Demming, you should check that the training image and test image dimensions are similar.

Now it's time to predict the identity of the face. Doing so takes only one line. Just call the predict() method on the recognizer object and pass it the face subarray. This method will return an ID number and a distance value.

The lower the distance value, the more likely the predicted face has been correctly identified. You can use the distance value as a threshold: all images that are predicted to be Captain Demming and score *at or below* the threshold will be positively identified as Captain Demming. All the others will be assigned to 'unknown'.

To apply the threshold, use an if statement ❷. If you're using your own training and test images, set the distance value to 1,000 the first time you run the program. Review the distance values for all the images in the test folder, both known and unknown. Find a threshold value below which all the faces are correctly identified as Captain Demming. This will be your discriminator going forward. For the images in the *demming_trainer* and *demming_tester* folders, the threshold distance should be 95.

Next, get the name for the image by using the predicted_id value as a key in the names dictionary. Print a message in the shell stating that the image has been identified and include the image filename, the name from the dictionary, and the distance value.

For the log, print a message indicating that name (in this case, Captain Demming) has been granted access to the lab and include the time using the datetime module ❸.

You'll want to keep a persistent file of people's comings and goings. Here's a neat trick for doing so: just write to a file using the print() function. Open the *lab_access_log.txt* file and include the a parameter for "append." This way, instead of overwriting the file for each new image, you'll add a new line at the bottom. Here's an example of the file contents:

```
Access granted to Demming at 2020-01-20 09:31:17.415802.
Access granted to Demming at 2020-01-20 09:31:19.556307.
Access granted to Demming at 2020-01-20 09:31:21.644038.
Access granted to Demming at 2020-01-20 09:31:23.691760.
--snip--
```

If the conditional is not met, set name to 'unknown' and print a message to that effect. Then draw a rectangle around the face and post the user's name using OpenCV's putText() method. Show the image for two seconds before destroying it.

Results

You can see some example results, from the 20 images in the *demming_tester* folder, in Figure 10-8. The predictor code correctly identified the eight images of Captain Demming with no false positives.

Figure 10-8: Demmings and non-Demmings

For the LBPH algorithm to be highly accurate, you need to use it under controlled conditions. Remember that by forcing the user to gain access through the laptop, you controlled their pose, the size of their face, the image resolution, and the lighting.

Summary

In this chapter, you got to work with OpenCV's local binary pattern histogram algorithm for recognizing human faces. With only a few lines of code, you produced a robust face recognizer that can easily handle variable lighting conditions. You also used the Standard Library's os.path.split() method to break apart directory paths and filenames to produce customized variable names.

Further Reading

"Local Binary Patterns Applied to Face Detection and Recognition" (Polytechnic University of Catalonia, 2010), by Laura María Sánchez López, is a thorough review of the LBPH approach. The PDF can be found online at sites such as *https://www.semanticscholar.org/*.

"Look at the LBP Histogram," on the AURlabCVsimulator site (*https://aurlabcvsimulator.readthedocs.io/en/latest/*), includes Python code that lets you visualize an LBPH image.

If you're a macOS or Linux user, be sure to check out Adam Geitgey's face_recognition library, a simple-to-use and highly accurate face recognition system that utilizes deep learning. You can find installation instructions and an overview at the Python Software Foundation site: *https://pypi.org/project/face_recognition/*.

"Machine Learning Is Fun! Part 4: Modern Face Recognition with Deep Learning" (Medium, 2016), by Adam Geitgey, is a short and enjoyable overview of modern face recognition using Python, OpenFace, and dlib.

"Liveness Detection with OpenCV" (PyImageSearch, 2019), by Adrian Rosebrock, is an online tutorial that teaches you how to protect your face recognition system against spoofing by fake faces, such as a photograph of Captain Demming held up to the webcam.

Cities and colleges around the world have begun banning facial recognition systems. Inventors have also gotten into the act, designing clothing that can confound the systems and protect your identity. "These Clothes Use Outlandish Designs to Trick Facial Recognition Software into Thinking You're Not Human" (Business Insider, 2020), by Aaron Holmes, and "How to Hack Your Face to Dodge the Rise of Facial Recognition Tech" (Wired, 2019), by Elise Thomas, review some recent practical—and impractical—solutions to the problem.

"OpenCV Age Detection with Deep Learning" (PyImageSearch, 2020) by Adrian Rosebrock, is an online tutorial for using OpenCV to predict a person's age from their photograph.

Challenge Project: Adding a Password and Video Capture

The *3_predict.py* program you wrote in Project 14 loops through a folder of photographs to perform face recognition. Rewrite the program so that it dynamically recognizes faces in the webcam's video stream. The face rectangle and name should appear in the video frame as they do on the folder images.

To start the program, have the user enter a password that you verify. If it's correct, add audio instructions telling the user to look at the camera. If the program positively identifies Captain Demming, use audio to announce that access is granted. Otherwise, play an audio message stating that access is denied.

If you need help with identifying the face from the video stream, see the *challenge_video_recognize.py* program in the appendix. Note that you may need to use a higher confidence value for the video frame than the value you used for the still photographs.

So that you can keep track of who has tried to enter the lab, save a single frame to the same folder as the *lab_access_log.txt* file. Use the logged results from datetime.now() as the filename so you can match the face to the access attempt. Note that you'll need to reformat the string returned from datetime.now() so that it only contains characters acceptable for filenames, as defined by your operating system.

Challenge Project: Look-Alikes and Twins

Use the code from Project 14 to compare celebrity look-alikes and twins. Train it with images from the internet and see whether you can fool the LBPH algorithm. Some pairings to consider are Scarlett Johansson and Amber Heard, Emma Watson and Kiernan Shipka, Liam Hemsworth and Karen Khachanov, Rob Lowe and Ian Somerhalder, Hilary Duff and Victoria Pedretti, Bryce Dallas Howard and Jessica Chastain, and Will Ferrell and Chad Smith.

For famous twins, look at astronaut twins Mark and Scott Kelly and celebrity twins Mary-Kate and Ashley Olsen.

Challenge Project: Time Machine

If you ever watch reruns of old shows, you'll encounter famous actors in their younger—sometimes *much* younger—days. Even though humans excel at face recognition, we may still struggle to identify a young Ian McKellen or Patrick Stewart. That's why sometimes it takes a certain inflection of voice or curious mannerism to send us scurrying to Google to check the cast members.

Face recognition algorithms are also prone to fail when identifying faces across time. To see how the LBPH algorithm performs under these conditions, use the code from Project 14 and train it on faces of yourself (or your relatives) at a certain age. Then test it with images over a range of ages.

11

CREATING AN INTERACTIVE ZOMBIE ESCAPE MAP

In 2010, *The Walking Dead* premiered on the AMC television channel. Set at the beginning of a zombie apocalypse, it told the story of a small group of survivors in the area of Atlanta, Georgia. The critically acclaimed show soon became a phenomenon, turning into the most watched series in cable television history, spawning a spin-off called *Fear the Walking Dead*, and starting an entirely new genre of television, the post-episode discussion show, with *Talking Dead*.

In this chapter, you'll play a quick-thinking data scientist who foresees the coming collapse of civilization. You'll prepare a map to help the *Walking Dead* survivors escape the crowded Atlanta metropolitan area for the more sparsely populated lands west of the Mississippi. In the process, you'll use the pandas library to load, analyze, and clean the data, and you'll use the bokeh and holoviews modules to plot the map.

According to scientists (yes, they've studied this), the key to surviving a zombie apocalypse is to live as far from a city as possible. In the United States, that means living in one of the large black areas shown in Figure 11-1. The brighter the lights, the greater the population, so if you want to avoid people, don't "go into the light."

Figure 11-1: Nighttime image of US city lights in 2012

Unfortunately for our *Walking Dead* survivors in Atlanta, they're a long way from the relative safety of the American West. They'll need to weave their way through a gauntlet of cities and towns, ideally passing through the least populated areas. Service station maps don't provide that population information, but the US census does. Before civilization collapses and the internet fails, you can download population density data onto your laptop and sort it out later using Python.

The best way to present this type of data is with a *choropleth map*, a visualization tool that uses colors or patterns to represent statistics about predefined geographical regions. You may be familiar with choropleth maps of US presidential election results, which color counties red for a Republican victory and blue for a Democratic one (Figure 11-2).

If the survivors had a choropleth map of population density that showed the number of people per square mile in each county, they could find the shortest, and theoretically safest, routes out of Atlanta and across the American South. Although you could get even higher-resolution data from the census, using its county-level data should be enough. *Walking Dead* zombie herds migrate as they get hungry, quickly rendering detailed statistics obsolete.

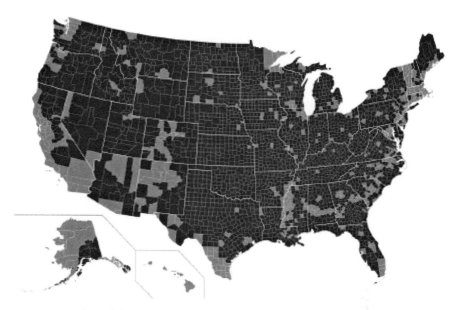

Figure 11-2: Choropleth map of the 2016 US presidential election results (light gray = Democrat, dark gray = Republican)

To determine the best routes through the counties, the survivors can use state highway maps like the ones found in service stations and welcome centers. These paper maps include county and parish outlines, making it easy to relate their network of cities and roads to a page-sized printout of the choropleth map.

THE OBJECTIVE

Create an interactive map of the conterminous United States (the 48 adjoining states) that displays population density by county.

The Strategy

Like all data visualization exercises, this task consists of the following basic steps: finding and cleaning the data, choosing the type of plot and the tool with which to show the data, preparing the data for plotting, and drawing the data.

Finding the data is easy in this case, as the US census population data is made readily available to the public. You still need to *clean* it, however, by finding and handling bogus data points, null values, and formatting issues. Ideally you would also verify the accuracy of the data, a difficult job that data scientists probably skip far too often. The data should at least pass a sanity check, something that may have to wait until the data is drawn. New York City should have a greater population density than Billings, Montana, for example.

Next, you must decide how you'll present the data. You'll use a map, but other options might include a bar chart or a table. Even more important is choosing the tool—in this case, the Python library—that you'll use to make the plot. The choice of tool can have a big impact on how you prepare the data and exactly what you end up showing.

Years ago, a fast-food company ran a commercial in which a customer claimed to like "a variety, but not too much of a variety." When it comes to visualization tools in Python, you can argue that there are too many choices, with too little to distinguish them: matplotlib, seaborn, plotly, bokeh, folium, altair, pygal, ggplot, holoviews, cartopy, geoplotlib, and built-in functions in pandas.

These various visualization libraries have their strengths and weaknesses, but since this project requires speed, you'll focus on the easy-to-use holoviews module, with a bokeh backend for plotting. This combination will allow you to produce an interactive choropleth map with only a few lines of code, and bokeh conveniently includes US state and county polygons in its sample data.

Once you've chosen your visualization tool, you must put the data in the format that the tool expects. You'll need to figure out how to fill in the county shapes, which you get from one file, with the population data from another file. This will involve a little reverse engineering using example code from the holoviews gallery. After that, you'll plot the map with bokeh.

Fortunately, data analysis with Python almost always relies on the Python Data Analysis Library (pandas). This module will let you load the census data, analyze it, and reformat it for use with holoviews and bokeh.

The Python Data Analysis Library

The open source pandas library is the most popular library available for performing data extraction, processing, and manipulation in Python. It contains data structures designed for working with common data sources, such as SQL relational databases and Excel spreadsheets. If you plan on being a data scientist in any form, you'll surely encounter pandas at some point.

The pandas library contains two primary data structures: series and dataframes. A *series* is a one-dimensional labeled array that can hold any type of data, such as integers, floats, strings, and so on. Because pandas is based on NumPy, a series object is basically two associated arrays (see the introduction to arrays on page 12 in Chapter 1 if you're new to arrays). One array contains the data point values, which can have any NumPy data type. The other array contains labels for each data point, called *indexes* (Table 11-1).

Table 11-1: A Series Object

Index	Value
0	25
1	432
2	–112
3	99

Unlike the indexes of Python list items, the indexes in a series don't have to be integers. In Table 11-2, the indexes are the names of people, and the values are their ages.

Table 11-2: A Series Object with Meaningful Labels

Index	Value
Javier	25
Carol	32
Lora	19
Sarah	29

As with a list or NumPy array, you can slice a series or select individual elements by specifying an index. You can manipulate the series many ways, such as filtering it, performing mathematical operations on it, and merging it with other series.

A *dataframe* is a more complex structure comprising two dimensions. It has a tabular structure similar to a spreadsheet, with columns, rows, and data (Table 11-3). You can think of it as an ordered collection of columns with two indexing arrays.

Table 11-3: A Dataframe Object

Index	Columns			
	Country	State	County	Population
0	USA	Alabama	Autauga	54,571
1	USA	Alabama	Baldwin	182,265
2	USA	Alabama	Barbour	27,457
3	USA	Alabama	Bibb	22,915

The first index, for the rows, works much like the index array in a series. The second keeps track of the series of labels, with each label representing a column header. Dataframes also resemble dictionaries; the column names form the keys, and the series of data in each column forms the values. This structure lets you easily manipulate dataframes.

Covering all the functionality in pandas would require a whole book, and you can find plenty online! We'll defer additional discussion until the code section, where we'll look at specific examples as we apply them.

The bokeh and holoviews Libraries

The bokeh module (*https://bokeh.org/*) is an open source interactive visualization library for modern web browsers. You can use it to construct elegant interactive graphics over large or streaming datasets. It renders its graphics using HTML and JavaScript, the predominant programming languages for creating interactive web pages.

The open source `holoviews` library (*http://holoviews.org/*) aims to make data analysis and visualization simple. With `holoviews`, instead of building a plot by making a direct call to a plotting library, such as `bokeh` or `matplotlib`, you first create an object describing your data, and the plots become automatic visual representations of this object.

The `holoviews` example gallery includes several choropleth maps visualized using bokeh (such as *http://holoviews.org/gallery/demos/bokeh/texas_choropleth _example.html*). Later, we'll use the unemployment rate example from this gallery to figure out how to present our population density data in a similar manner.

Installing pandas, bokeh, and holoviews

If you worked through the project in Chapter 1, you already have `pandas` and `NumPy` installed. If not, see the instructions in "Installing the Python Libraries" on page 6.

One option for installing `holoviews`, along with latest version of all the recommended packages for working with the module on Linux, Windows, or macOS, is to use Anaconda.

```
conda install -c pyviz holoviews bokeh
```

This installation method includes the default `matplotlib` plotting library backend, the more interactive `bokeh` plotting library backend, and the Jupyter/IPython Notebook.

You can install a similar set of packages using `pip`.

```
pip install 'holoviews[recommended]'
```

Additional minimal installation options are available through `pip`, assuming you already have `bokeh` installed. You can find these and other installation instructions at *http://holoviews.org/install.html* and *http://holoviews .org/user_guide/Installing_and_Configuring.html*.

Accessing the County, State, Unemployment, and Population Data

The `bokeh` library comes with data files for the state and county outlines and the 2009 US unemployment data per county. As mentioned, you'll use the unemployment data to determine how to format the population data, which comes from the 2010 census.

To download the `bokeh` sample data, connect to the internet, open a Python shell, and enter the following:

```
>>> import bokeh
>>> import bokeh.sampledata
>>> bokeh.sampledata.download()
Creating C:\Users\lee_v\.bokeh directory
Creating C:\Users\lee_v\.bokeh\data directory
Using data directory: C:\Users\lee_v\.bokeh\data
```

As you can see, the program will tell you where on your machine it's putting the data so that bokeh can automatically find it. Your path will differ from mine. For more on downloading the sample data, see *https://docs.bokeh .org/en/latest/docs/reference/sampledata.html*.

Look for *US_Counties.csv* and *unemployment09.csv* in the folder of downloaded files. These plaintext files use the popular *comma-separated values* (CSV) format, in which each line represents a data record with multiple fields separated by commas. (Good luck saying "CSV" right if you regularly shop at a CVS pharmacy!)

The unemployment file is instructive of the plight of the data scientist. If you open it, you'll see that there are no column names describing the data (Figure 11-3), though it's possible to guess what most of the fields represent. We'll deal with this later.

	A	B	C	D	E	F	G	H	I
1	CN010010	1	1	Autauga County, AL	2009	23,288	21,025	2,263	9.7
2	PA011000	1	3	Baldwin County, AL	2009	81,706	74,238	7,468	9.1
3	CN010050	1	5	Barbour County, AL	2009	9,703	8,401	1,302	13.4
4	CN010070	1	7	Bibb County, AL	2009	8,475	7,453	1,022	12.1
5	CN010090	1	9	Blount County, AL	2009	25,306	22,789	2,517	9.9
6	CN010110	1	11	Bullock County, AL	2009	3,527	2,948	579	16.4

Figure 11-3: The first few rows of unemployment09.csv

If you open the US counties file, you'll see lots of columns, but at least they have headers (Figure 11-4). Your challenge will be to relate the unemployment data in Figure 11-3 to the geographical data in Figure 11-4 so that you can do the same later with the census data.

	A	B	C	D	E	F	G	H	I	J	K	L
1	County Name	State-County	state abbr	State Abbr.	geometry	value	GEO_ID	GEO_ID2	Geographic Name	STATE num	COUNTY num	FIPS formula
2	Autauga	AL-Autauga	al	AL	<Polygon>	126.4	05000US01001	1001	Autauga County, Alabama	1	1	1001
3	Baldwin	AL-Baldwin	al	AL	<Polygon>	486.1	05000US01003	1003	Baldwin County, Alabama	1	3	1003
4	Barbour	AL-Barbour	al	AL	<Polygon>	583.3	05000US01005	1005	Barbour County, Alabama	1	5	1005
5	Bibb	AL-Bibb	al	AL	<Polygon>	569.3	05000US01007	1007	Bibb County, Alabama	1	7	1007
6	Blount	AL-Blount	al	AL	<Polygon>	893	05000US01009	1009	Blount County, Alabama	1	9	1009

Figure 11-4: The first few rows of US_Counties.csv

You can find the population data, *census_data_popl_2010.csv*, in the *Chapter_11* folder, downloadable from the book's website. This file, originally named *DEC_10_SF1_GCTPH1.US05PR_with_ann.csv*, came from the American FactFinder website. By the time this book is published, the US government will have migrated the census data to a new site called *https:// data.census.gov* (see *https://www.census.gov/data/what-is-data-census-gov.html*).

If you look at the top of the census file, you'll see lots of columns with two header rows (Figure 11-5). You're interested in column M, titled *Density per square mile of land area – Population*.

	A	B	C	D	E	F
1	GEO.id	GEO.id2	GEO.display-label	GCT_STUB.target-geo-id	GCT_STUB.target-geo-id2	GCT_STUB.display-label
2	Id	Id2	Geography	Target Geo Id	Target Geo Id2	Geographic area
3	0100000US		United States	0100000US		United States
4	0100000US		United States	0400000US01	1	United States - Alabama
5	0100000US		United States	0500000US01001	1001	United States - Alabama ·
6	0100000US		United States	0500000US01003	1003	United States - Alabama ·

	G	H	I	J	K	L	M	N
1	GCT_STUB.display-label	HD01	HD02	SUBHD0301	SUBHD0302	SUBHD0303	SUBHD0401	SUBHD0402
2	Geographic area	Populat	Housin	Area in squa	Area in squa	Area in squa	Density per	Density per
3	United States	308745!	131704;	3796742.23	264836.79	3531905.43	87.4	37.3
4	Alabama	477973(	217185;	52420.07	1774.74	50645.33	94.4	42.9
5	Autauga County	54571	22135	604.39	9.95	594.44	91.8	37.2
6	Baldwin County	182265	104061	2027.31	437.53	1589.78	114.6	65.5

Figure 11-5: The first few rows of census_data_popl_2010.csv

At this point, you have all the Python libraries and data files you need to generate a population density choropleth map *in theory*. Before you can write the code, however, you need to know how you're going to link the population data to the geographical data so that you can place the correct county data in the correct county shape.

Hacking holoviews

Learning to adapt existing code for your own use is a valuable skill for a data scientist. This may require a bit of reverse engineering. Because open source software is free, it's sometimes poorly documented, so you have to figure out how it works on your own. Let's take a moment and apply this skill to our current problem.

In previous chapters, we took advantage of the gallery examples provided by open source modules such as turtle and matplotlib. The holoviews library also has a gallery (*http://holoviews.org/gallery/index.html*), and it includes Texas Choropleth Example, a choropleth map of the Texas unemployment rate in 2009 (Figure 11-6).

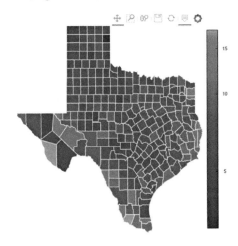

Figure 11-6: Choropleth map of the 2009 Texas unemployment rate from the holoviews gallery

Listing 11-1 contains the code provided by holoviews for this map. You'll build your project based on this example, but to do so, you'll have to address two main differences. First, you plan to plot population density rather than unemployment rate. Second, you want a map of the conterminous United States, not just Texas.

texas_ choropleth_ example.html

```
import holoviews as hv
from holoviews import opts
hv.extension('bokeh')
❶ from bokeh.sampledata.us_counties import data as counties
from bokeh.sampledata.unemployment import data as unemployment

counties = [dict(county, ❷Unemployment=unemployment[cid])
                for cid, county in counties.items()
                ❸ if county["state"] == "tx"]

choropleth = hv.Polygons(counties, ['lons', 'lats'],
                         [('detailed name', 'County'), 'Unemployment'])

choropleth.opts(opts.Polygons(logz=True,
                              tools=['hover'],
                              xaxis=None, yaxis=None,
                              show_grid=False,
                              show_frame=False,
                              width=500, height=500,
                              color_index='Unemployment',
                              colorbar=True, toolbar='above',
                              line_color='white'))
```

Listing 11-1: holoviews gallery code for generating the Texas choropleth

The code imports the data from the bokeh sample data ❶. You'll need to know the format and content of both the unemployment and counties variables. The unemployment rate is accessed later using the unemployment variable and an index or key of cid, which may stand for "county ID" ❷. The program selects Texas, rather than the whole United States, based on a conditional using a state code ❸.

Let's investigate this in the Python shell.

```
>>> from bokeh.sampledata.unemployment import data as unemployment
❶ >>> type(unemployment)
<class 'dict'>
❷ >>> first_2 = {k: unemployment[k] for k in list(unemployment)[:2]}
>>> for k in first_2:
        print(f"{k} : {first_2[k]}")
❸ (1, 1) : 9.7
(1, 3) : 9.1
>>>
>>> for k in first_2:
        for item in k:
            print(f"{item}: {type(item)}")
```

```
❹ 1: <class 'int'>
   1: <class 'int'>
   1: <class 'int'>
   3: <class 'int'>
```

Start by importing the bokeh sample data using the syntax from the gallery example. Next, use the type() built-in function to check the data type of the unemployment variable ❶. You'll see that it's a dictionary.

Now, use dictionary comprehension to make a new dictionary comprising the first two lines in unemployment ❷. Print the results, and you'll see that the keys are tuples and the values are numbers, presumably the unemployment rate in percent ❸. Check the data type for the numbers in the key. They're integers rather than strings ❹.

Compare the output at ❸ to the first two rows in the CSV file in Figure 11-3. The first number in the key tuple, presumably a state code, comes from column B. The second number in the tuple, presumably a county code, comes from column C. The unemployment rate is obviously stored in column I.

Now compare the contents of unemployment to Figure 11-4, representing the county data. The *STATE num* (column J) and *COUNTY num* (column K) obviously hold the components of the key tuple.

So far so good, but if you look at the population data file in Figure 11-5, you won't find a state or county code to direct into a tuple. There are numbers in column E, however, that match those in the last column of the county data, labeled *FIPS formula* in Figure 11-4. These FIPS numbers seem to relate to the state and county codes.

As it turns out, a *Federal Information Processing Series (FIPS)* code is basically a ZIP code for a county. The FIPS code is a five-digit numeric code assigned to each county by the National Institute of Standards and Technology. The first two digits represent the county's state, and the final three digits represent the county (Table 11-4).

Table 11-4: Identifying US Counties Using a FIPS Code

US County	State Code	County Code	FIPS
Baldwin County, AL	01	003	1003
Johnson County, IA	19	103	19103

Congratulations, you now know how to link the US census data to the county shapes in the bokeh sample data. It's time to write the final code!

The Choropleth Code

The *choropleth.py* program includes code for both cleaning the data and plotting the choropleth map. You can find a copy of the code, along with the population data, in the *Chapter_11* folder downloadable from the book's website at *https://nostarch.com/real-world-python/*.

Importing Modules and Data and Constructing a Dataframe

Listing 11-2 imports modules and the bokeh county sample data that includes coordinates for all the US county polygons. It also loads and creates a dataframe object to represent the population data. Then it begins the process of cleaning and preparing the data for use with the county data.

choropleth.py, part 1

```
from os.path import abspath
import webbrowser
import pandas as pd
import holoviews as hv
from holoviews import opts
❶ hv.extension('bokeh')
from bokeh.sampledata.us_counties import data as counties

❷ df = pd.read_csv('census_data_popl_2010.csv', encoding="ISO-8859-1")

df = pd.DataFrame(df,
                  columns=
                  ['Target Geo Id2',
                  'Geographic area.1',
                  'Density per square mile of land area - Population'])

df.rename(columns =
          {'Target Geo Id2':'fips',
           'Geographic area.1': 'County',
           'Density per square mile of land area - Population':'Density'},
          inplace = True)

print(f"\nInitial popl data:\n {df.head()}")
print(f"Shape of df = {df.shape}\n")
```

Listing 11-2: Importing modules and data, creating a dataframe, and renaming columns

Start by importing abspath from the operating system library. You'll use this to find the absolute path to the choropleth map HTML file after it's created. Then import the webbrowser module so you can launch the HTML file. You need this because the holoviews library is designed to work with a Jupyter Notebook and won't automatically display the map without some help.

Next, import pandas and repeat the holoviews imports from the gallery example in Listing 11-1. Note that you must specify bokeh as the holoviews extension, or *backend* ❶. This is because holoviews can work with other plotting libraries, such as matplotlib, and needs to know which one to use.

You brought in the geographical data with the imports. Now load the population data using pandas. This module includes a set of input/output API functions to facilitate reading and writing data. These *readers* and *writers* address major formats such as comma-separated values (read_csv, to_csv), Excel (read_excel, to_excel), Structured Query Language (read_sql, to_sql), HyperText Markup Language (read_html, to_html), and more. In this project, you'll work with the CSV format.

In most cases, you can read CSV files without specifying the character encoding.

```
df = pd.read_csv('census_data_popl_2010.csv')
```

In this case, however, you'll get the following error:

```
UnicodeDecodeError: 'utf-8' codec can't decode byte 0xf1 in position 31:
invalid continuation byte
```

That's because the file contains characters encoded with Latin-1, also known as ISO-8859-1, rather than the default UTF-8 encoding. Adding the encoding argument will fix the problem ❷.

Now, turn the population data file into a tabular dataframe by calling the DataFrame() constructor. You don't need all the columns in the original file, so pass the names of the column you want to keep to the constructor. These represent columns E, G, and M in Figure 11-5, or the FIPS code, county name (without the state name), and population density, respectively.

Next, use the rename() dataframe method to make the column labels shorter and more meaningful. Call them *fips*, *County*, and *Density*.

Finish the listing by printing the first few rows of the dataframe using the head() method and by printing the shape of the dataframe using its shape attribute. By default, the head() method prints the first five rows. If you want to see more rows, you can pass it the number as an argument, such as head(20). You should see the following output in the shell:

```
Initial popl data:
      fips          County  Density
0      NaN   United States    87.4
1      1.0         Alabama    94.4
2   1001.0  Autauga County    91.8
3   1003.0  Baldwin County   114.6
4   1005.0  Barbour County    31.0
Shape of df = (3274, 3)
```

Notice that the first two rows (rows 0 and 1) are not useful. In fact, you can glean from this output that each state will have a row for its name, which you'll want to delete. You can also see from the shape attribute that there are 3,274 rows in the dataframe.

Removing Extraneous State Name Rows and Preparing the State and County Codes

Listing 11-3 removes all rows whose FIPS code is less than or equal to 100. These are header rows that indicate where a new state begins. It then creates new columns for the state and county codes, which it derives from the existing column of FIPS codes. You'll use these later to select the proper county outline from the bokeh sample data.

```
df = df[df['fips'] > 100]
print(f"Popl data with non-county rows removed:\n {df.head()}")
print(f"Shape of df = {df.shape}\n")

❶ df['state_id'] = (df['fips'] // 1000).astype('int64')
df['cid'] = (df['fips'] % 1000).astype('int64')
print(f"Popl data with new ID columns:\n {df.head()}")
print(f"Shape of df = {df.shape}\n")
print("df info:")
❷ print(df.info())

print("\nPopl data at row 500:")
❸ print(df.loc[500])
```

Listing 11-3: Removing extraneous rows and preparing the state and county codes

To display the population density data in the county polygons, you need to turn it into a dictionary where the keys are a tuple of the state code and county code and the values are the density data. But as you saw previously, the population data does not include separate columns for the state and county codes; it has only the FIPS codes. So, you'll need to split out the state and county components.

First, get rid of all the noncounty rows. If you look at the previous shell output (or rows 3 and 4 in Figure 11-5), you'll see that these rows do not include a four- or five-digit FIPS code. You can thus use the fips column to make a new dataframe, still named df, that preserves only rows with a fips value greater than 100. To check that this worked, repeat the printout from the previous listing, as shown here:

```
Popl data with non-county rows removed:
      fips          County  Density
2   1001.0  Autauga County     91.8
3   1003.0  Baldwin County    114.6
4   1005.0  Barbour County     31.0
5   1007.0     Bibb County     36.8
6   1009.0   Blount County     88.9
Shape of df = (3221, 3)
```

The two "bad" rows at the top of the dataframe are now gone, and based on the shape attribute, you've lost a total of 53 rows. These represent the header rows for the 50 states, United States, District of Columbia (DC), and Puerto Rico. Note that DC has a FIPS code of 11001 and Puerto Rico uses a state code of 72 to go with the three-digit county code for its 78 municipalities. You'll keep DC but remove Puerto Rico later.

Next, create columns for state and county code numbers. Name the first new column state_id ❶. Dividing by 1,000 using floor division (//) returns the quotient with the digits after the decimal point removed. Since the last three numbers of the FIPS code are reserved for county codes, this leaves you with the state code.

Although `//` returns an integer, the new dataframe column uses the float datatype by default. But our analysis of the bokeh sample data indicated that it used integers for these codes in the key tuples. Convert the column to the integer datatype using the pandas `astype()` method and pass it `'int64'`.

Now, make a new column for the county code. Name it `cid` so it will match the terminology used in the holoviews choropleth example. Since you're after the last three digits in the FIPS code, use the modulo operator (`%`). This returns the remainder from the division of the first argument to the second. Convert this column to the integer datatype as in the previous line.

Print the output again, only this time call the `info()` method on the dataframe ❷. This method returns a concise summary of the dataframe, including datatypes and memory usage.

```
Popl data with new ID columns:
      fips          County  Density  state_id  cid
2  1001.0  Autauga County     91.8         1    1
3  1003.0  Baldwin County    114.6         1    3
4  1005.0  Barbour County     31.0         1    5
5  1007.0     Bibb County     36.8         1    7
6  1009.0   Blount County     88.9         1    9
Shape of df = (3221, 5)

df info:
<class 'pandas.core.frame.DataFrame'>
Int64Index: 3221 entries, 2 to 3273
Data columns (total 5 columns):
fips        3221 non-null float64
County      3221 non-null object
Density     3221 non-null float64
state_id    3221 non-null int64
cid         3221 non-null int64
dtypes: float64(2), int64(2), object(1)
memory usage: 151.0+ KB
None
```

As you can see from the columns and information summary, the state_id and cid numbers are integer values.

The state codes in the first five rows are all single-digit numbers, but it's possible for state codes to have double digits, as well. Take the time to check the state codes of later rows. You can do this by calling the `loc()` method on the dataframe and passing it a high row number ❸. This will let you check double-digit state codes.

```
Popl data at row 500:
fips                  13207
County       Monroe County
Density                66.8
state_id                 13
cid                     207
Name: 500, dtype: object
```

The `fips`, `state_id`, and `cid` all look reasonable. This completes the data preparation. The next step is to turn this data into a dictionary that `holoviews` can use to make the choropleth map.

Preparing the Data for Display

Listing 11-4 converts the state and county IDs and the density data into separate lists. It then recombines them into a dictionary with the same format as the `unemployment` dictionary used in the `holoviews` gallery example. It also lists the states and territories to exclude from the map and makes a list of the data needed to plot the choropleth map.

```
state_ids = df.state_id.tolist()
cids = df.cid.tolist()
den = df.Density.tolist()

tuple_list = tuple(zip(state_ids, cids))
popl_dens_dict = dict(zip(tuple_list, den))

EXCLUDED = ('ak', 'hi', 'pr', 'gu', 'vi', 'mp', 'as')

counties = [dict(county, Density=popl_dens_dict[cid])
            for cid, county in counties.items()
            if county["state"] not in EXCLUDED]
```

Listing 11-4: Preparing the population data for plotting

Earlier, we looked at the `unemployment` variable in the `holoviews` gallery example and found that it was a dictionary. Tuples of the state and county codes served as the keys, and the unemployment rates served as the values, as follows:

```
(1, 1) : 9.7
(1, 3) : 9.1
--snip--
```

To create a similar dictionary for the population data, first use the pandas `tolist()` method to create separate lists of the dataframe's `state_id`, `cid`, and `Density` columns. Then, use the built-in `zip()` function to merge the state and county code lists as tuple pairs. Create the final dictionary, `popl_dens_dict`, by zipping this new `tuple_list` with the density list. (The name `tuple_list` is misleading; technically, it's a `tuple_tuple`.) That's it for the data preparation.

The *Walking Dead* survivors will be lucky to get out of Atlanta. Let's forget about them reaching Alaska. Make a tuple, named `EXCLUDED`, of states and territories that are in the `bokeh` county data but aren't part of the conterminous United States. These include Alaska, Hawaii, Puerto Rico, Guam, Virgin Islands, Northern Mariana Islands, and American Samoa. To reduce typing, you can use the abbreviations provided as a column in the county dataset (see Figure 11-4).

Creating an Interactive Zombie Escape Map   **259**

Next, as in the holoviews example, make a dictionary and put it in a list named counties. Here's where you add the population density data. Link it to the proper county using the cid county identifier number. Use a conditional to apply the EXCLUDED tuple.

If you print the first index in this list, you'll get the (truncated) output that follows:

```
[{'name': 'Autauga', 'detailed name': 'Autauga County, Alabama', 'state':
'al', 'lats': [32.4757, 32.46599, 32.45054, 32.44245, 32.43993, 32.42573,
32.42417, --snip-- -86.41231, -86.41234, -86.4122, -86.41212, -86.41197,
-86.41197, -86.41187], 'Density': 91.8}]
```

The Density key-value pair now replaces the unemployment rate pair used in the holoviews gallery example. Next up, plotting the map!

Plotting the Choropleth Map

Listing 11-5 creates the choropleth map, saves it as an *.html* file, and opens it with the webbrowser.

choropleth.py,
part 4

```
choropleth = hv.Polygons(counties,
                         ['lons', 'lats'],
                         [('detailed name', 'County'), 'Density'])
```

❶ ```
choropleth.opts(opts.Polygons(logz=True,
                     tools=['hover'],
                     xaxis=None, yaxis=None,
                     show_grid=False, show_frame=False,
                     width=1100, height=700,
                     colorbar=True, toolbar='above',
                     color_index='Density', cmap='Greys', line_color=None,
                     title='2010 Population Density per Square Mile of Land Area'
                     ))
```

❷ ```
hv.save(choropleth, 'choropleth.html', backend='bokeh')
url = abspath('choropleth.html')
webbrowser.open(url)
```

Listing 11-5: Creating and plotting the choropleth map

According to the holoviews documentation, the Polygons() class creates a contiguous filled area in a 2D space as a list of polygon geometries. Name a variable choropleth and pass Polygons() the counties variable and the dictionary keys, including the lons and lats used to draw the county polygons. Also pass it the county names and population density keys. The holoviews hover tool will use this tuple, ('detailed name', 'County'), to show you the full county name, such as County: Claiborne County, Mississippi, when you move the cursor around the map (Figure 11-7).

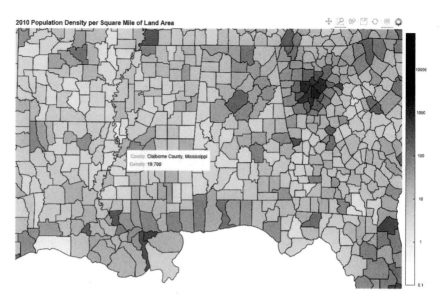

2010 Population Density per Square Mile of Land Area

County: Claiborne County, Mississippi
Density: 19.700

10000
1000
100
10
1
0.1

Figure 11-7: Choropleth map with the hover feature active

Next, set the options for the map ❶. First, permit use of a logarithmic color bar by setting the logz argument to True.

The holoviews window will come with a set of default tools such as pan, zoom, save, refresh, and so on (see the upper-right corner of Figure 11-7). Use the tools argument to add the hover feature to this list. This allows you query the map and get both the county name and detailed information on the population density.

You're not making a standard plot with an annotated *x*-axis and *y*-axis, so set these to None. Likewise, don't show a grid or frame around the map.

Set the width and height of the map in pixels. You may want to adjust this for your monitor. Next set colorbar to True and place the toolbar at the top of the display.

Since you want to color the counties based on population density, set the color_index argument to Density, which represents the values in popl_dens _dict. For the fill colors, use the Greys cmap. If you want to use brighter colors, you can find a list of available colormaps at *http://build.holoviews.org/user_guide /Colormaps.html*. Be sure to choose one with "bokeh" in the name. Finish the color scheme by selecting a line color for the county outlines. Good choices for a gray colormap are None, 'white', or 'black'.

Complete the options by adding a title. The choropleth map is now ready for plotting.

To save your map in the current directory, use the holoviews save() method and pass it the choropleth variable, a file name with the *.html* extension, and the name of the plotting backend being used ❷. As mentioned previously, holoviews is designed for use with a Jupyter Notebook. If you want the map to automatically pop up on your browser, first assign the full path to the saved map to a url variable. Then use the webbrowser module to open url and display the map (Figure 11-8).

Figure 11-8: The 2010 population density choropleth map. Lighter colors represent lower population density

You can use the toolbar at the top of the map to pan, zoom (using a box or lasso), save, refresh, or hover. The hover tool, shown in Figure 11-7, will help you find the least populated counties in places where the map shading makes the difference hard to distinguish visually.

NOTE *The Box Zoom tool permits a quick view of a rectangular area but may stretch or squeeze the map axes. To preserve the map's aspect ratio when zooming, use a combination of the Wheel Zoom and Pan tools.*

Planning the Escape

The Chisos Mountains, an extinct supervolcano in Big Bend National Park, might be one of the best places on Earth to ride out a zombie apocalypse. Remote and fortress-like in appearance (Figure 11-9), the mountains tower 4,000 feet above the surrounding desert plain, reaching a maximum elevation of almost 8,000 feet. At their heart lies a natural basin with park facilities, including a lodge, cabins, store, and restaurant. Fish and game are abundant in the area, desert springs provide fresh water, and the banks of the Rio Grande are suitable for farming.

Figure 11-9: The Chisos Mountains of west Texas (left) with 3D relief map representation (right)

With your choropleth map, you can quickly plan a route to this natural fortress far, far away. But first, you need to escape Atlanta. The shortest route out of the metropolitan area is a narrow passage squeezed between the cities of Birmingham and Montgomery in Alabama (Figure 11-10). You can skirt the next big city, Jackson, Mississippi, by going either north or south. To choose the best route, however, you need to look farther ahead.

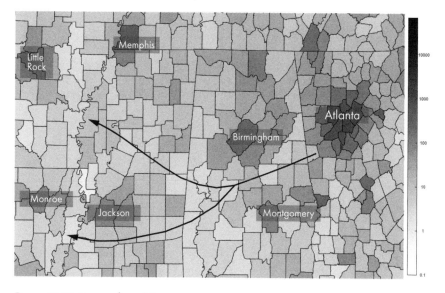

Figure 11-10: Escape from Atlanta

The southerly route around Jackson is shorter but forces the survivors to pass over the highly developed I-35 corridor, anchored by San Antonio in the south and Dallas–Fort Worth (DFW) in the north (Figure 11-11). This creates a potentially dangerous choke point at Hill County, Texas (circled in Figure 11-11).

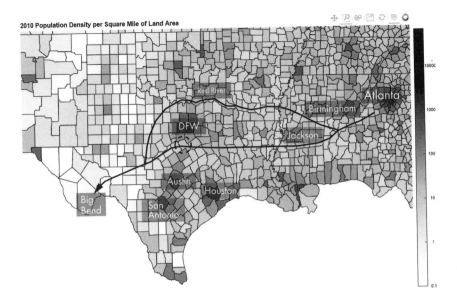

Figure 11-11: The way west

Alternatively, the northerly route through the Red River Valley, between Oklahoma and Texas, would be longer but safer, especially if you took advantage of the navigable river. Once west of Fort Worth, the survivors could cross the river and turn south to salvation.

This type of planning would be even simpler if holoviews provided a slider tool that allowed you to interactively alter the color bar. For example, you could filter out or change the shading of counties by simply dragging your cursor up and down the legend. This would make it easier to find connected routes through the lowest population counties.

Unfortunately, a slider tool isn't one of the holowviews window options. Since you know pandas, though, that won't stop you. Simply add the following snippet of code after the line that prints the information at location 500:

```
df.loc[df.Density >= 65, ['Density']] = 1000
```

This will change the population density values in the dataframe, setting those greater than or equal to 65 to a constant value of 1000. Run the program again, and you'll get the plot in Figure 11-12. With the new values, the San Antonio–Austin–Dallas barrier becomes more apparent, as does the relative safety of the Red River Valley that forms the northern border of east Texas.

You may be wondering, where did the survivors go in the TV show? They went nowhere! They spent the first four seasons in the vicinity of Atlanta, first camping at Stone Mountain and then holed up in a prison near the fictional town of Woodbury (Figure 11-13).

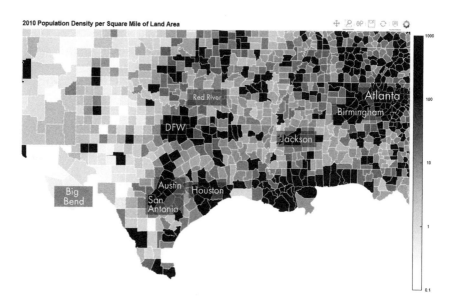

Figure 11-12: Counties with more than 65 people per square mile shaded black

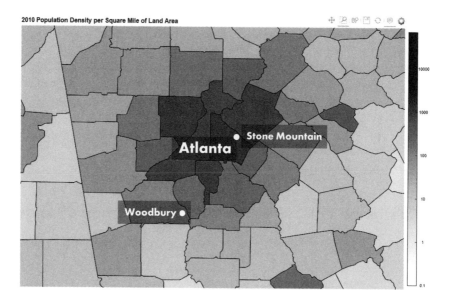

Figure 11-13: Location of Stone Mountain and the fictional town of Woodbury

Stone Mountain is less than 20 miles from downtown Atlanta and in DeKalb County, with 2,586 people per square mile. Woodbury (the real town of Senoia) is only 35 miles from downtown Atlanta and on the border of Coweta County, with 289 people per square mile, and Fayette County, with 549 people per square mile. No wonder these guys had so much trouble. If only there had been a data scientist in the group.

Summary

In this chapter, you got to work with the Python Data Analysis Library (pandas) and the bokeh and holoviews visualization modules. In the process, you did some real-world data wrangling to clean and link data from different sources.

Further Reading

"If the Zombie Apocalypse Happens, Scientists Say You Should Run for the Hills" (Business Insider, 2017), by Kevin Loria, describes the application of standard disease models to infection rates in a theoretical zombie outbreak.

"What to Consider When Creating Choropleth Maps" (*Chartable*, 2018), by Lisa Charlotte Rost, provides useful guidelines for making choropleth maps. You can find it at *https://blog.datawrapper.de/choroplethmaps/*.

"Muddy America: Color Balancing the Election Map—Infographic" (STEM Lounge, 2019) by Larry Weru, demonstrates ways to increase the useful detail in choropleth maps, using the iconic red-blue United States election map as an example.

Python Data Science Handbook: Essential Tools for Working with Data (O'Reilly Media, 2016), by Jake VanderPlas, is a thorough reference for important Python data science tools, including pandas.

Beneath the Window: Early Ranch Life in the Big Bend Country (Iron Mountain Press, 2003), by Patricia Wilson Clothier, is an engaging recollection of growing up in the early 20th century on a vast ranch in the Big Bend country of Texas, before it became a national park. It provides insight into how apocalypse survivors might deal with life in the harsh country.

Game Theory: Real Tips for SURVIVING a Zombie Apocalypse (7 Days to Die) (The Game Theorists, 2016) is a video on the best place *in the world* to escape a zombie apocalypse. Unlike *The Walking Dead*, the video assumes that the zombie virus can be transmitted by mosquitoes and ticks, and it selects the location with this in mind. It's available online.

Challenge Project: Mapping US Population Change

The US government will release population data from the 2020 census in 2021. However, less accurate, intercensal population estimates for 2019 are currently available. Use one of these, along with the 2010 data from Project 15, to generate a new choropleth map that captures population change, by county, over that time period.

Hint: you can subtract columns in pandas dataframes to generate the difference data, as demonstrated in the toy example that follows. The 2020 population values represent dummy data.

```
>>> import pandas as pd
>>>
>>> # Generate example population data by county:
>>> pop_2010 = {'county': ['Autauga', 'Baldwin', 'Barbour', 'Bibb'],
            'popl': [54571, 182265, 27457, 22915]}
```

```
>>> pop_2020 = {'county': ['Autauga', 'Baldwin', 'Barbour', 'Bibb'],
                'popl': [52910, 258321, 29073, 29881]}
>>>
>>> df_2010 = pd.DataFrame(pop_2010)
>>> df_2020 = pd.DataFrame(pop_2020)
>>> df_diff = df_2020.copy()   # Copy the 2020 dataframe to a new df
>>> df_diff['diff'] = df_diff['popl'].sub(df_2010['popl'])   # Subtract popl columns
>>> print(df_diff.loc[:4, ['county', 'diff']])
    county    diff
0  Autauga   -1661
1  Baldwin   76056
2  Barbour    1616
3     Bibb    6966
```

12

ARE WE LIVING IN A COMPUTER SIMULATION?

In 2003, the philosopher Nick Bostrom postulated that we live in a computer simulation run by our advanced, possibly posthuman, descendants. Today, many scientists and big thinkers, including Neil DeGrasse Tyson and Elon Musk, believe there's a good chance this *simulation hypothesis* is true. It certainly explains why mathematics so elegantly describes nature, why observers seem to influence quantum events, and why we appear to be alone in the universe.

Even stranger, *you* could be the only real thing in this simulation. Perhaps you're a brain in a vat, immersing yourself in a historical simulation. For computational efficiency, the simulation might render only those things with which you currently interact. When you go inside and close your door, the world outside might turn off like a refrigerator light. How would you really know one way or the other?

Scientists take this hypothesis seriously, holding debates and publishing papers on how we might devise some test to prove it. In this chapter, you'll attempt to answer the question using an approach proposed by physicists: you'll build a simple simulated world and then analyze it for clues that might give the simulation away. In doing so, you'll work through this project backward, writing the code before coming up with the problem-solving strategy. You'll find that even the simplest model can provide profound insights on the nature of our existence.

Project #16: Life, the Universe, and Yertle's Pond

The ability to simulate reality isn't just a far-off dream. Physicists have used the world's most powerful supercomputers to accomplish this feat, simulating subatomic particle behavior at a scale of a few femtometers (10^{-15} m). Although the simulation represents only a tiny piece of the cosmos, it's indistinguishable from what we understand to be reality.

But don't worry, you won't need a supercomputer or a degree in physics to solve this problem. All you need is the turtle module, a drawing program designed for kids. You used turtle to simulate the Apollo 8 mission in Chapter 6. Here, you'll use it to understand one of the foundational features of computer models. You'll then apply that knowledge to devise the same basic strategy that physicists plan to apply to the simulation hypothesis.

THE OBJECTIVE

Identify a feature of a computer simulation that might be detectable by those being simulated.

The Pond Simulation Code

The *pond_sim.py* code creates a turtle-based simulation of a pond that includes a mud island, a floating log, and a snapping turtle named Yertle. Yertle will swim out to the log, swim back, and then swim out again. You can download the code from the book's website at *https://nostarch.com /real-world-python/*.

The turtle module ships with Python, so you don't have to install anything. For an overview of the module, see "Using the turtle Module" on page 127.

Importing turtle, Setting Up the Screen, and Drawing the Island

Listing 12-1 imports turtle, sets up a screen object to use as a pond, and draws a mud island for Yertle to survey his domain.

pond_sim.py, part 1
```
import turtle

pond = turtle.Screen()
pond.setup(600, 400)
```

```
pond.bgcolor('light blue')
pond.title("Yertle's Pond")

mud = turtle.Turtle('circle')
mud.shapesize(stretch_wid=5, stretch_len=5, outline=None)
mud.pencolor('tan')
mud.fillcolor('tan')
```

Listing 12-1: Importing the turtle module and drawing a pond and mud island

After importing the turtle module, assign a screen object to a variable named pond. Use the turtle setup() method to set the screen size, in pixels, and then color the background light blue. You can find tables of turtle colors and their names on multiple sites, such as *https://trinket.io/docs/colors*. Finish the pond by providing a title for the screen.

Next, make a circular mud island for Yertle to sunbathe on. Use the Turtle() class to instantiate a turtle object named mud. Although turtle comes with a method for drawing circles, it's easier here to just pass the constructor the 'circle' argument, which produces a circular turtle object. This circle shape is too small to make much of an island, however, so use the shapesize() method to stretch it out. Finish the island by setting its outline and fill colors to tan.

Drawing the Log, a Knothole, and Yertle

Listing 12-2 completes the program by drawing the log, complete with knothole and Yertle the turtle. It then moves Yertle so that he can leave his island to check out the log.

pond_sim.py, part 2
```
SIDE = 80
ANGLE = 90
log = turtle.Turtle()
log.hideturtle()
log.pencolor('peru')
log.fillcolor('peru')
log.speed(0)
❶ log.penup()
log.setpos(215, -30)
log.lt(45)
log.begin_fill()
❷ for _ in range(2):
    log.fd(SIDE)
    log.lt(ANGLE)
    log.fd(SIDE / 4)
    log.lt(ANGLE)
log.end_fill()

knot = turtle.Turtle()
knot.hideturtle()
knot.speed(0)
knot.penup()
knot.setpos(245, 5)
knot.begin_fill()
```

```
knot.circle(5)
knot.end_fill()

yertle = turtle.Turtle('turtle')
yertle.color('green')
yertle.speed(1)  # Slowest.
yertle.fd(200)
yertle.lt(180)
yertle.fd(200)
❸ yertle.rt(176)
yertle.fd(200)
```

Listing 12-2: Drawing a log and a turtle and then moving the turtle around

You'll draw a rectangle to represent the log, so start by assigning two constants, SIDE and ANGLE. The first represents the length of the log, in pixels; the second is the angle, in degrees, by which you'll turn the turtle at each corner of the rectangle.

By default, all turtles initially appear at the center of the screen, at coordinates (0, 0). Since you'll place your log off to the side, after you instantiate the log object, use the hideturtle() method to make it invisible. This way, you don't have to watch it fly across the screen to get to its final position.

Color the log brown, using peru for the log color. Then set the object's speed to the fastest setting (oddly, 0). This way, you won't have to watch it slowly draw on the screen. And so you don't see the path it takes from the screen's center to its edge, pick up the drawing pen using the penup() method ❶.

Use the setpos() method—for *set position*—to place the log near the right edge of the screen. Then turn the object left by 45 degrees and call the begin_fill() method.

You can save a few lines of code by drawing the rectangle using a for loop ❷. You'll loop twice, drawing two sides of the rectangle with each loop. Make the log's width 20 pixels by dividing SIDE by 4. After the loop, call end_fill() to color the log brown.

Give the log some character by adding a knothole, represented by a knot turtle. To draw the knothole, call the circle() method and pass it 5, for a radius of five pixels. Note that you don't need to specify a fill color as black is the default.

Finally, end the program by drawing Yertle, the king of all he surveys. Yertle is an old turtle, so set his drawing speed to the slowest setting of 1. Have him swim out and inspect the log and then turn around and swim back. Yertle is a touch senile, and he forgets what he just did. So, have him swim back out—only this time, angle his course so that he's no longer swimming due east ❸. Run the program, and you should get the results shown in Figure 12-1.

Look carefully at this figure. Despite the simplicity of the simulation, it contains powerful insights into whether we, like Yertle, dwell in a computer simulation.

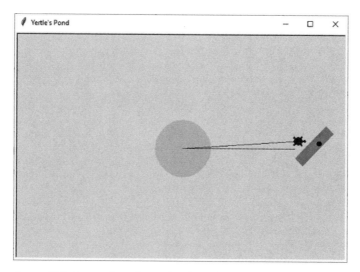

Figure 12-1: Screenshot of completed simulation

Implications of the Pond Simulation

Because of finite computational resources, all computer simulations require a framework of some type on which to "hang" their model of reality. Whether it's called a grid, a lattice, a mesh, a matrix, or whatever, it provides a way to both distribute objects in 2D or 3D space and assign them a property, such as mass, temperature, color, or something else.

The turtle module uses the pixels in your monitor as its coordinate system, as well as to store properties. The pixel locations define the shapes, such as the log's outline, and the pixel color property helps differentiate one shape from another.

Pixels form an *orthogonal* pattern, which means the rows and columns of pixels intersect at right angles. Although individual pixels are square and too small to easily see, you can use the turtle module's dot() method to generate a facsimile, as in the following snippet:

```
>>> import turtle
>>>
>>> t = turtle.Turtle()
>>> t.hideturtle()
>>> t.penup()
>>>
>>> def dotfunc(x, y):
        t.setpos(x, y)
        for _ in range(10):
                t.dot()
                t.fd(10)

>>> for i in range(0, 100, 10):
        dotfunc(0, -i)
```

This produces the pattern in Figure 12-2.

Figure 12-2: Orthogonal grid of black dots
representing the centers of square pixels

In the turtle world, pixels are true atoms: indivisible. A line can't be shorter than one pixel. Movement can occur only as integers of pixels (though you can input float values without raising an error). The smallest object possible is one pixel in size.

An implication of this is that the simulation's grid determines the smallest feature you can observe. Since we can observe incredibly small subatomic particles, our grid, assuming we're a simulation, must be incredibly fine. This leads many scientists to seriously doubt the simulation conjecture, since it would require a staggering amount of computer memory. Still, who knows what our distant descendants, or aliens, are capable of?

Besides setting a limit on the size of objects, a simulation grid might force a preferred orientation, or *anisotropy*, on the fabric of the cosmos. Anisotropy is the directional dependence of a material, such as the way wood splits more easily along its grain rather than across it. If you look closely at Yertle's paths in the turtle simulation (Figure 12-3), you can see evidence of anisotropy. His upper, slightly angled path zigzags, while the lower, east-west path is perfectly straight.

Figure 12-3: The angled versus straight path

Drawing a nonorthogonal line on an orthogonal grid isn't pretty. But there's more involved than just aesthetics. Moving along the *x* or *y* direction requires only integer addition or subtraction (Figure 12-4, left). Moving at an angle requires trigonometry to calculate the partial movement in the *x* and *y* directions (Figure 12-4, right).

For a computer, mathematical calculations equal work, so we can surmise that moving at an angle takes more energy. By timing the two calculations in Figure 12-4, we can get a relative measure of this difference in energy.

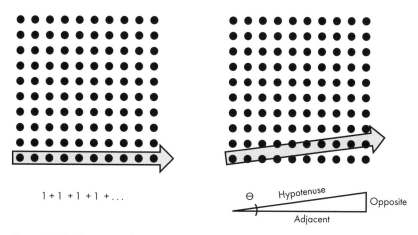

$1 + 1 + 1 + 1 + \ldots$

Figure 12-4: Movement along rows or columns (left) requires simpler arithmetic than moving across them (right)

Measuring the Cost of Crossing the Lattice

To time the difference between drawing a line diagonally across a pixel grid and drawing the line along it, you need to draw two lines of equal length. But remember, turtle works only with integers. You need to find an angle for which all sides of a triangle—the opposite, adjacent, and hypotenuse in Figure 12-4—are integers. This way, you'll know that your angled line is the same length as your straight line.

To find these angles, you can use a *Pythagorean triple*, a set of positive integers *a*, *b*, and *c* that fit the right triangle rule $a^2 + b^2 = c^2$. The best-known triple is 3-4-5, but you'll want a longer line, to ensure that the runtime of the drawing function isn't less than the measurement precision of your computer's clock. Fortunately, you can find other, larger triples online. The triplet 62-960-962 is a good choice, as it's long but will still fit in a turtle screen.

The Line Comparison Code

To compare the cost of drawing a diagonal line to the cost of drawing a straight one, Listing 12-3 uses turtle to draw the two lines. The first line is parallel to the *x*-axis (that is, east-west), and the second line is at a shallow angle to the *x*-axis. You can figure out the correct degree of the angle using trigonometry; in this case, it's 3.695220532 degrees. The listing draws these lines many times using a for loop and records the time it takes to draw each one using the built-in time module. The final comparison uses the averages of these runs.

You need to use averages because your central processing unit (CPU) is constantly running multiple processes. The operating system schedules these processes behind the scenes, executing one while delaying another until a resource, such as input/output, becomes available. Consequently, it's difficult to record the *absolute* runtime of a given function. Calculating the average time of many runs compensates for this.

You can download the code, *line_compare.py*, from the book's website.

line_compare.py

```
from time import perf_counter
import statistics
import turtle

turtle.setup(1200, 600)
screen = turtle.Screen()

ANGLES = (0, 3.695220532)  # In degrees.
NUM_RUNS = 20
SPEED = 0
for angle in ANGLES:
 ❶  times = []
    for _ in range(NUM_RUNS):
        line = turtle.Turtle()
        line.speed(SPEED)
        line.hideturtle()
        line.penup()
        line.lt(angle)
        line.setpos(-470, 0)
        line.pendown()
        line.showturtle()
     ❷  start_time = perf_counter()
        line.fd(962)
        end_time = perf_counter()
        times.append(end_time - start_time)

    line_ave = statistics.mean(times)
    print("Angle {} degrees: average time for {} runs at speed {} = {:.5f}"
          .format(angle, NUM_RUNS, SPEED, line_ave))
```

Listing 12-3: Drawing a straight line and an angled line and recording the runtimes for each

Start by importing perf_counter—short for *performance counter*—from the time module. This function returns the float value of time in seconds. It gives you a more precise answer than time.clock(), which it replaces as of Python 3.8.

Next, import the statistics module to help you calculate the average of many simulation runs. Then import turtle and set up the turtle screen. You can customize the screen for your monitor, but remember, you need to be able to see a line 962 pixels long.

Now, assign some key values for the simulation. Put the angles for a straight line and a diagonal line in a tuple named ANGLES and then assign a variable to hold the number of times to run the for loop and the speed at which to draw the line.

Start looping through the angles in the ANGLES tuple. Create an empty list to hold the time measurements ❶ before setting up a turtle object, as you've done before. Rotate the turtle object left by the angle amount and then use setpos() to move it to the far-left side of the screen.

Move the turtle forward by 962 pixels, sandwiching this command between calls to `perf_counter()` to time the movement ❷. Subtract the end time from the start time and append the result to the `times` list.

Finish by using the `statistics.mean()` function to find the average runtime for each line. Print the results to five decimal places. After the program runs, the turtle screen should look like Figure 12-5.

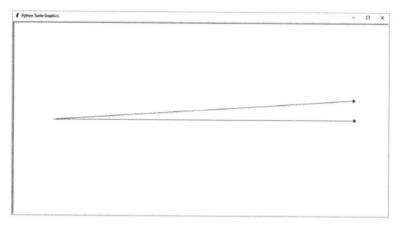

Figure 12-5: Completed turtle screen for line_compare.py

Because you used a Pythagorean triple, the angled line truly ends on a pixel. It doesn't just snap to the nearest pixel. Consequently, you can be confident that the straight and angled lines have the same length and that you're comparing apples to apples when it comes to the timing measurements.

Results

If you draw each line 500 times and then compare the results, you should see that it takes roughly 2.4 times as long to draw the angled line as the straight line.

```
Angle 0 degrees: average time for 500 runs at speed 0 = 0.06492
Angle 3.695220532 degrees: average time for 500 runs at speed 0 = 0.15691
```

Your times will likely differ slightly, as they're affected by other programs you may have running concurrently on your computer. As noted previously, CPU scheduling will manage all these processes so that your system is fast, efficient, and fair.

If you repeat the exercise for 1,000 runs, you should get similar results. (If you decide to do so, you'll want to get yourself a cup of coffee and some of that good pie.) The angled line will take about 2.7 times as long to draw.

```
Angle 0 degrees: average time for 1000 runs at speed 0 = 0.10911
Angle 3.695220532 degrees: average time for 1000 runs at speed 0 = 0.29681
```

You've been running a short function at a high drawing speed. If you're worried that turtle performs optimizations to achieve speed at the expense of accuracy, you can slow it down and rerun the program. With the drawing speed set to normal (speed = 6), the angled line takes about 2.6 times as long to draw, close to the outcome with the fastest speed.

```
Angle 0 degrees: average time for 500 runs at speed 6 = 1.12522
Angle 3.695220532 degrees: average time for 500 runs at speed 6 = 2.90180
```

Clearly, moving across the pixel grid requires more work than moving along it.

The Strategy

The goal of this project was to identify a way for simulated beings, perhaps us, to find evidence of the simulation. At this point, we know at least two things. First, if we're living in a simulation, the grid is extremely small, as we can observe subatomic particles. Second, if these small particles cross the simulation's grid at an angle, we should expect to find computational resistance that translates into something measurable. This resistance might look like a loss of energy, a scattering of particles, a reduction in velocity, or something similar.

In 2012, physicists Silas R. Beane, from the University of Bonn, and Zohreh Davoudi and Martin J. Savage, from the University of Washington, published a paper arguing exactly this point. According to the authors, if the laws of physics, which appear continuous, are superimposed on a discrete grid, the grid spacing might impose a limitation on physical processes.

They proposed investigating this by observing *ultra-high energy cosmic rays (UHECRs)*. UHECRs are the fastest particles in the universe, and they are affected by increasingly smaller features as they get more energetic. But there's a limit to how much energy these particles can have. Known as the GZK cutoff and confirmed by experiments in 2007, this limit is consistent with the kind of boundary a simulation grid might cause. Such a boundary should also cause UHECRs to travel preferentially along the grid's axes and scatter particles that try to cross it.

Not surprisingly, there are many potential obstacles to this approach. UHECRs are rare, and anomalous behavior might not be obvious. If the spacing of the grid is significantly smaller than 10^{-12} femtometers, we probably can't detect it. There may not even *be* a grid, at least as we understand it, as the technology in use may far exceed our own. And, as the philosopher Preston Greene pointed out in 2019, there may be a moral obstacle to the entire project. If we live in a simulation, our discovery of it may trigger its end!

Summary

From a coding standpoint, building Yertle's simulated world was simple. But a big part of coding is solving problems, and the small amount of work you did had major implications. No, we didn't make the leap to cosmic rays, but we started the right conversation. The basic premise that a computer simulation requires a grid that could imprint observable signatures on the universe is an idea that transcends nitty-gritty details.

In the book *Harry Potter and the Deathly Hallows*, Harry asks the wizard Dumbledore, "Tell me one last thing. Is this real? Or has this been happening inside my head?" Dumbledore replies, "Of course it is happening inside your head, Harry, but why on Earth should that mean that it is not real?"

Even if our world isn't located at the "fundamental level of reality," as Nick Bostrom postulates, you can still take pleasure in your ability to solve problems such as this. As Descartes might've said, had he lived today, "I code, therefore I am." Onward!

Further Reading

"Are We Living in a Simulated Universe? Here's What Scientists Say" (NBC News, 2019), by Dan Falk, provides an overview of the simulation hypothesis.

"Neil deGrasse Tyson Says 'It's Very Likely' the Universe Is a Simulation" (ExtremeTech, 2016), by Graham Templeton, is an article with an embedded video of the Isaac Asimov Memorial Debate, hosted by astrophysicist Neil deGrasse Tyson, that addresses the possibility that we're living in a simulation.

"Are We Living in a Computer Simulation? Let's Not Find Out" (*New York Times*, 2019), by Preston Greene, presents a philosophical argument against investigating the simulation hypothesis.

"We Are Not Living in a Simulation. Probably." (Fast Company, 2018), by Glenn McDonald, argues that the universe is too big and too detailed to be simulated computationally.

Moving On

There's never enough time in life to do all the things we want, and that goes double for writing a book. The challenge projects that follow represent the ghosts of chapters not yet written. There was no time to finish these (or in some cases, even start them), but you might have better luck. As always, the book provides no solutions for challenge projects—not that you'll need them.

This is the real world, baby, and you're ready for it.

Challenge Project: Finding a Safe Space

The award-winning 1970 novel *Ringworld* introduced the world to the Pierson's puppeteer, a sentient and highly advanced alien herbivore. Being herd animals, puppeteers were extremely cowardly and cautious. When they realized that the core of the Milky Way had exploded and the radiation would reach them in 20,000 years, they started fleeing the galaxy immediately!

In this project, you're part of a 29th-century diplomatic team assigned to the puppeteer ambassador. Your job is to select a state, within the conterminous United States, that they'll find suitably safe for the puppeteer embassy. You'll need to screen each state for natural hazards, such as earthquakes, volcanoes, tornadoes, and hurricanes, and present the ambassador with a map summarizing the results. Don't worry that the data you'll use is hundreds of years out-of-date; just pretend it's current to the year 2850 CE.

You can find earthquake data at *https://earthquake.usgs.gov/earthquakes/feed/v1.0/csv.php/*. Use dots to plot the epicenters of those quakes that are 6.0 or greater in magnitude.

You can post the tornado data as the average number per year per state (see *https://www.ncdc.noaa.gov/climate-information/extreme-events/us-tornado-climatology*). Use a choropleth format like you did in Chapter 11.

You can find a listing of dangerous volcanoes in Table 2 of the 2018 Update to the U.S. Geological Survey National Volcanic Threat Assessment (*https://pubs.usgs.gov/sir/2018/5140/sir20185140.pdf*). Represent these as dots on the map, but assign them a different color or shape than the earthquake data. Also, ignore the ashfall from Yellowstone. Assume the experts monitoring this supervolcano can predict an eruption soon enough for the ambassador to safely flee the planet.

To find hurricane tracks, visit the National Oceanic and Atmospheric Administration site (*https://coast.noaa.gov/digitalcoast/data/*) and search for "Historical Hurricane Tracks." Download and post the Category 4 and higher storm segments on the map.

Try to think like a puppeteer and use the final composite map to choose a candidate state for the embassy. You might have to ignore a tornado or two. America is a dangerous place!

Challenge Project: Here Comes the Sun

In 2018, 13-year-old Georgia Hutchinson from Woodside, California, won $25,000 at the Broadcom Masters nationwide science, technology, engineering, and mathematics (STEM) competition for middle-school students. Her entry, "Designing a Data-Driven Dual-Axis Solar Tracker," will make solar panels cheaper and more efficient by eliminating the need for costly light sensors.

This new sun tracker is based on the premise that we already know the location of the sun at any moment from any given point on Earth. It uses public data from the National Oceanic and Atmospheric Administration to continuously determine the sun's position and tilt the solar panels for maximum power production.

Write a Python program that calculates the sun's position based on a location of your choosing. To get started, check out the Wikipedia page "Position of the Sun" (*https://en.wikipedia.org/wiki/Position_of_the_Sun*).

Challenge Project: Seeing Through a Dog's Eyes

Use your knowledge of computer vision to write a Python program that takes an image and simulates what a dog would see. To get started, check out *https://www.akc.org/expert-advice/health/are-dogs-color-blind/* and *https://dog-vision.andraspeter.com/*.

Challenge Project: Customized Word Search

Boy, does your Granny love doing word searches! For her birthday, use Python to design and print her customized word searches using family names, vintage TV shows like *Matlock* and *Columbo*, or the common names of her prescription drugs. Allow the words to print horizontally, vertically, and diagonally.

Challenge Project: Simplifying a Celebration Slideshow

Your spouse, sibling, parent, best friend, or whoever is having a celebration dinner, and you're in charge of the slideshow. You have tons of pictures in the cloud, many featuring the honoree, but the filenames just list the date and time at which they were taken, providing no clue as to the contents. It looks like you'll spend your Saturday sifting through them all.

But wait, didn't you learn about face recognition in that book *Real-World Python*? All you really need to do is find a few training images and do a bit of coding.

First, pick someone in your personal digital photo collection to represent the guest of honor. Next, write a Python program that searches through your folders, finds photos containing this person, and copies the photos into a special folder for your review. When training, be sure to include face profiles as well as frontal views, and include a profile Haar cascade when detecting faces.

Challenge Project: What a Tangled Web We Weave

Use Python and the turtle module to simulate a spider building a web. For some guidance on web construction, see *https://www.brisbaneinsects.com/brisbane_weavers/index.htm* and *http://recursiveprocess.com/mathprojects/index.php/2015/06/09/spider-webs-creepy-or-cool/*.

Challenge Project: Go Tell It on the Mountain

"What's the closest mountain to Houston, Texas?" This seemingly straightforward question, asked on Quora, isn't easy to answer. For one thing, you need to consider mountains in Mexico, as well as those in the United States. For another, there's no universally accepted definition of a mountain.

To make this somewhat easier, use one of the UN Environmental Program's definitions of *mountainous terrain*. Find prominences with an elevation of at least 2,500 m (8,200 feet) and consider them mountains. Calculate their distance from the center of Houston to find the closest.

PRACTICE PROJECT SOLUTIONS

This appendix contains solutions to the practice projects in each chapter. Digital versions are available on the book's website at *https://nostarch.com/real-world-python/*.

Chapter 2: Attributing Authorship with Stylometry

Hunting the Hound with Dispersion

*practice_hound
_dispersion.py*

```
"""Use NLP (nltk) to make dispersion plot."""
import nltk
import file_loader

corpus = file_loader.text_to_string('hound.txt')
tokens = nltk.word_tokenize(corpus)
tokens = nltk.Text(tokens)  # NLTK wrapper for automatic text analysis.
```

```
dispersion = tokens.dispersion_plot(['Holmes',
                                      'Watson',
                                      'Mortimer',
                                      'Henry',
                                      'Barrymore',
                                      'Stapleton',
                                      'Selden',
                                      'hound'])
```

Punctuation Heatmap

practice_heatmap
_semicolon.py

```
"""Make a heatmap of punctuation."""
import math
from string import punctuation
import nltk
import numpy as np
import matplotlib.pyplot as plt
from matplotlib.colors import ListedColormap
import seaborn as sns

# Install seaborn using: pip install seaborn.

PUNCT_SET = set(punctuation)

def main():
    # Load text files into dictionary by author.
    strings_by_author = dict()
    strings_by_author['doyle'] = text_to_string('hound.txt')
    strings_by_author['wells'] = text_to_string('war.txt')
    strings_by_author['unknown'] = text_to_string('lost.txt')

    # Tokenize text strings preserving only punctuation marks.
    punct_by_author = make_punct_dict(strings_by_author)

    # Convert punctuation marks to numerical values and plot heatmaps.
    plt.ion()
    for author in punct_by_author:
        heat = convert_punct_to_number(punct_by_author, author)
        arr = np.array((heat[:6561])) # trim to largest size for square array
        arr_reshaped = arr.reshape(int(math.sqrt(len(arr))),
                                   int(math.sqrt(len(arr))))
        fig, ax = plt.subplots(figsize=(7, 7))
        sns.heatmap(arr_reshaped,
                    cmap=ListedColormap(['blue', 'yellow']),
                    square=True,
                    ax=ax)
        ax.set_title('Heatmap Semicolons {}'.format(author))
    plt.show()
```

```
def text_to_string(filename):
    """Read a text file and return a string."""
    with open(filename) as infile:
        return infile.read()

def make_punct_dict(strings_by_author):
    """Return dictionary of tokenized punctuation by corpus by author."""
    punct_by_author = dict()
    for author in strings_by_author:
        tokens = nltk.word_tokenize(strings_by_author[author])
        punct_by_author[author] = ([token for token in tokens
                                    if token in PUNCT_SET])
        print("Number punctuation marks in {} = {}"
              .format(author, len(punct_by_author[author])))
    return punct_by_author

def convert_punct_to_number(punct_by_author, author):
    """Return list of punctuation marks converted to numerical values."""
    heat_vals = []
    for char in punct_by_author[author]:
        if char == ';':
            value = 1
        else:
            value = 2
        heat_vals.append(value)
    return heat_vals

if __name__ == '__main__':
    main()
```

Chapter 4: Sending Super-Secret Messages with a Book Cipher

Charting the Characters

*practice
_barchart.py*
```
"""Plot barchart of characters in text file."""
import sys
import os
import operator
from collections import Counter
import matplotlib.pyplot as plt

def load_file(infile):
    """Read and return text file as string of lowercase characters."""
    with open(infile) as f:
        text = f.read().lower()
    return text
```

```
def main():
    infile = 'lost.txt'
    if not os.path.exists(infile):
        print("File {} not found. Terminating.".format(infile),
              file=sys.stderr)
        sys.exit(1)

    text = load_file(infile)

    # Make bar chart of characters in text and their frequency.
    char_freq = Counter(text)
    char_freq_sorted = sorted(char_freq.items(),
                              key=operator.itemgetter(1), reverse=True)
    x, y = zip(*char_freq_sorted)  # * unpacks iterable.
    fig, ax = plt.subplots()
    ax.bar(x, y)
    fig.show()

if __name__ == '__main__':
    main()
```

Sending Secrets the WWII Way

*practice_WWII
_words.py*

```
"""Book code using the novel The Lost World

For words not in book, spell-out with first letter of words.
Flag 'first letter mode' by bracketing between alternating
'a a' and 'the the'.

credit: Eric T. Mortenson
"""
import sys
import os
import random
import string
from collections import defaultdict, Counter

def main():
    message = input("Enter plaintext or ciphertext: ")
    process = input("Enter 'encrypt' or 'decrypt': ")
    shift = int(input("Shift value (1-365) = "))
    infile = input("Enter filename with extension: ")

    if not os.path.exists(infile):
        print("File {} not found. Terminating.".format(infile), file=sys.stderr)
        sys.exit(1)
    word_list = load_file(infile)
    word_dict = make_dict(word_list, shift)
    letter_dict = make_letter_dict(word_list)

    if process == 'encrypt':
        ciphertext = encrypt(message, word_dict, letter_dict)
        count = Counter(ciphertext)
```

```python
        encryptedWordList = []
        for number in ciphertext:
            encryptedWordList.append(word_list[number - shift])

        print("\nencrypted word list = \n {} \n"
              .format(' '.join(encryptedWordList)))
        print("encrypted ciphertext = \n {}\n".format(ciphertext))

        # Check the encryption by decrypting the ciphertext.
        print("decrypted plaintext = ")
        singleFirstCheck = False
        for cnt, i in enumerate(ciphertext):
            if word_list[ciphertext[cnt]-shift] == 'a' and \
               word_list[ciphertext[cnt+1]-shift] == 'a':
                continue
            if word_list[ciphertext[cnt]-shift] == 'a' and \
               word_list[ciphertext[cnt-1]-shift] == 'a':
                singleFirstCheck = True
                continue
            if singleFirstCheck == True and cnt<len(ciphertext)-1 and \
               word_list[ciphertext[cnt]-shift] == 'the' and \
                          word_list[ciphertext[cnt+1]-shift] == 'the':
                continue
            if singleFirstCheck == True and \
               word_list[ciphertext[cnt]-shift] == 'the' and \
                          word_list[ciphertext[cnt-1]-shift] == 'the':
                singleFirstCheck = False
                print(' ', end='', flush=True)
                continue
            if singleFirstCheck == True:
                print(word_list[i - shift][0], end = '', flush=True)
            if singleFirstCheck == False:
                print(word_list[i - shift], end=' ', flush=True)

    elif process == 'decrypt':
        plaintext = decrypt(message, word_list, shift)
        print("\ndecrypted plaintext = \n {}".format(plaintext))

def load_file(infile):
    """Read and return text file as a list of lowercase words."""
    with open(infile, encoding='utf-8') as file:
        words = [word.lower() for line in file for word in line.split()]
        words_no_punct = ["".join(char for char in word if char not in \
                                   string.punctuation) for word in words]
    return words_no_punct

def make_dict(word_list, shift):
    """Return dictionary of characters as keys and shifted indexes as values."""
    word_dict = defaultdict(list)
    for index, word in enumerate(word_list):
        word_dict[word].append(index + shift)
    return word_dict

def make_letter_dict(word_list):
    firstLetterDict = defaultdict(list)
```

```
        for word in word_list:
            if len(word) > 0:
                if word[0].isalpha():
                    firstLetterDict[word[0]].append(word)
        return firstLetterDict

    def encrypt(message, word_dict, letter_dict):
        """Return list of indexes representing characters in a message."""
        encrypted = []
        # remove punctuation from message words
        messageWords = message.lower().split()
        messageWordsNoPunct = ["".join(char for char in word if char not in \
                                       string.punctuation) for word in messageWords]
        for word in messageWordsNoPunct:
            if len(word_dict[word]) > 1:
                index = random.choice(word_dict[word])
            elif len(word_dict[word]) == 1:  # Random.choice fails if only 1 choice.
                index = word_dict[word][0]
            elif len(word_dict[word]) == 0:  # Word not in word_dict.
                encrypted.append(random.choice(word_dict['a']))
                encrypted.append(random.choice(word_dict['a']))

                for letter in word:
                    if letter not in letter_dict.keys():
                        print('\nLetter {} not in letter-to-word dictionary.'
                              .format(letter), file=sys.stderr)
                        continue
                    if len(letter_dict[letter])>1:
                        newWord =random.choice(letter_dict[letter])
                    else:
                        newWord = letter_dict[letter][0]
                    if len(word_dict[newWord])>1:
                        index = random.choice(word_dict[newWord])
                    else:
                        index = word_dict[newWord][0]
                    encrypted.append(index)

                encrypted.append(random.choice(word_dict['the']))
                encrypted.append(random.choice(word_dict['the']))
                continue
            encrypted.append(index)
        return encrypted

    def decrypt(message, word_list, shift):
        """Decrypt ciphertext string and return plaintext word string.

        This shows how plaintext looks before extracting first letters.
        """
        plaintextList = []
        indexes = [s.replace(',', '').replace('[', '').replace(']', '')
                   for s in message.split()]
        for count, i in enumerate(indexes):
            plaintextList.append(word_list[int(i) - shift])
        return ' '.join(plaintextList)
```

```
def check_for_fail(ciphertext):
    """Return True if ciphertext contains any duplicate keys."""
    check = [k for k, v in Counter(ciphertext).items() if v > 1]
    if len(check) > 0:
        print(check)
        return True


if __name__ == '__main__':
    main()
```

Chapter 5: Finding Pluto

Plotting the Orbital Path

*practice_orbital
_path.py*

```
import os
from pathlib import Path
import cv2 as cv

PAD = 5  # Ignore pixels this distance from edge

def find_transient(image, diff_image, pad):
    """Takes image, difference image, and pad value in pixels and returns
        boolean and location of maxVal in difference image excluding an edge
        rind. Draws circle around maxVal on image."""
    transient = False
    height, width = diff_image.shape
    cv.rectangle(image, (PAD, PAD), (width - PAD, height - PAD), 255, 1)
    minVal, maxVal, minLoc, maxLoc = cv.minMaxLoc(diff_image)
    if pad < maxLoc[0] < width - pad and pad < maxLoc[1] < height - pad:
        cv.circle(image, maxLoc, 10, 255, 0)
        transient = True
    return transient, maxLoc

def main():
    night1_files = sorted(os.listdir('night_1_registered_transients'))
    night2_files = sorted(os.listdir('night_2'))
    path1 = Path.cwd() / 'night_1_registered_transients'
    path2 = Path.cwd() / 'night_2'
    path3 = Path.cwd() / 'night_1_2_transients'

    # Images should all be the same size and similar exposures.
    for i, _ in enumerate(night1_files[:-1]):  # Leave off negative image
        img1 = cv.imread(str(path1 / night1_files[i]), cv.IMREAD_GRAYSCALE)
        img2 = cv.imread(str(path2 / night2_files[i]), cv.IMREAD_GRAYSCALE)

        # Get absolute difference between images.
        diff_imgs1_2 = cv.absdiff(img1, img2)
        cv.imshow('Difference', diff_imgs1_2)
        cv.waitKey(2000)
```

```
# Copy difference image and find and circle brightest pixel.
temp = diff_imgs1_2.copy()
transient1, transient_loc1 = find_transient(img1, temp, PAD)

# Draw black circle on temporary image to obliterate brightest spot.
cv.circle(temp, transient_loc1, 10, 0, -1)

# Get location of new brightest pixel and circle it on input image.
transient2, transient_loc2 = find_transient(img1, temp, PAD)

if transient1 or transient2:
    print('\nTRANSIENT DETECTED between {} and {}\n'
          .format(night1_files[i], night2_files[i]))
    font = cv.FONT_HERSHEY_COMPLEX_SMALL
    cv.putText(img1, night1_files[i], (10, 25),
               font, 1, (255, 255, 255), 1, cv.LINE_AA)
    cv.putText(img1, night2_files[i], (10, 55),
               font, 1, (255, 255, 255), 1, cv.LINE_AA)
    if transient1 and transient2:
        cv.line(img1, transient_loc1, transient_loc2, (255, 255, 255),
                1, lineType=cv.LINE_AA)

    blended = cv.addWeighted(img1, 1, diff_imgs1_2, 1, 0)
    cv.imshow('Surveyed', blended)
    cv.waitKey(2500)  # Keeps window open 2.5 seconds.

    out_filename = '{}_DECTECTED.png'.format(night1_files[i][:-4])
    cv.imwrite(str(path3 / out_filename), blended)  # Will overwrite!

else:
    print('\nNo transient detected between {} and {}\n'
          .format(night1_files[i], night2_files[i]))

if __name__ == '__main__':
    main()
```

What's the Difference?

This practice project uses two programs, *practice_montage_aligner.py* and *practice_montage_difference_finder.py*. The programs should be run in the order presented.

practice_montage_aligner.py

practice_montage _aligner.py

```
import numpy as np
import cv2 as cv

MIN_NUM_KEYPOINT_MATCHES = 150

img1 = cv.imread('montage_left.JPG', cv.IMREAD_COLOR)  # queryImage
img2 = cv.imread('montage_right.JPG', cv.IMREAD_COLOR) # trainImage
```

```
img1 = cv.cvtColor(img1, cv.COLOR_BGR2GRAY)  # Convert to grayscale.
img2 = cv.cvtColor(img2, cv.COLOR_BGR2GRAY)

orb = cv.ORB_create(nfeatures=700)

# Find the keypoints and descriptions with ORB.
kp1, desc1 = orb.detectAndCompute(img1, None)
kp2, desc2 = orb.detectAndCompute(img2, None)

# Find keypoint matches using Brute Force Matcher.
bf = cv.BFMatcher(cv.NORM_HAMMING, crossCheck=True)
matches = bf.match(desc1, desc2, None)

# Sort matches in ascending order of distance.
matches = sorted(matches, key=lambda x: x.distance)

# Draw best matches.
img3 = cv.drawMatches(img1, kp1, img2, kp2,
                      matches[:MIN_NUM_KEYPOINT_MATCHES],
                      None)

cv.namedWindow('Matches', cv.WINDOW_NORMAL)
img3_resize = cv.resize(img3, (699, 700))
cv.imshow('Matches', img3_resize)
cv.waitKey(7000)  # Keeps window open 7 seconds.
cv.destroyWindow('Matches')

# Keep only best matches.
best_matches = matches[:MIN_NUM_KEYPOINT_MATCHES]

if len(best_matches) >= MIN_NUM_KEYPOINT_MATCHES:
    src_pts = np.zeros((len(best_matches), 2), dtype=np.float32)
    dst_pts = np.zeros((len(best_matches), 2), dtype=np.float32)

    for i, match in enumerate(best_matches):
        src_pts[i, :] = kp1[match.queryIdx].pt
        dst_pts[i, :] = kp2[match.trainIdx].pt

    M, mask = cv.findHomography(src_pts, dst_pts, cv.RANSAC)

    # Get dimensions of image 2.
    height, width = img2.shape
    img1_warped = cv.warpPerspective(img1, M, (width, height))

    cv.imwrite('montage_left_registered.JPG', img1_warped)
    cv.imwrite('montage_right_gray.JPG', img2)

else:
    print("\n{}\n".format('WARNING: Number of keypoint matches < 10!'))
```

practice_montage_difference_finder.py

```python
import cv2 as cv

filename1 = 'montage_left.JPG'
filename2 = 'montage_right_gray.JPG'

img1 = cv.imread(filename1, cv.IMREAD_GRAYSCALE)
img2 = cv.imread(filename2, cv.IMREAD_GRAYSCALE)

# Absolute difference between image 2 & 3:
diff_imgs1_2 = cv.absdiff(img1, img2)

cv.namedWindow('Difference', cv.WINDOW_NORMAL)
diff_imgs1_2_resize = cv.resize(diff_imgs1_2, (699, 700))
cv.imshow('Difference', diff_imgs1_2_resize)

crop_diff = diff_imgs1_2[10:2795, 10:2445]  # x, y, w, h = 10, 10, 2790, 2440

# Blur to remove extraneous noise.
blurred = cv.GaussianBlur(crop_diff, (5, 5), 0)

(minVal, maxVal, minLoc, maxLoc2) = cv.minMaxLoc(blurred)
cv.circle(img2, maxLoc2, 100, 0, 3)
x, y = int(img2.shape[1]/4), int(img2.shape[0]/4)
img2_resize = cv.resize(img2, (x, y))
cv.imshow('Change', img2_resize)
```

Chapter 6: Winning the Moon Race with Apollo 8

Simulating a Search Pattern

```python
import time
import random
import turtle

SA_X = 600  # Search area width.
SA_Y = 480  # Search area height.
TRACK_SPACING = 40  # Distance between search tracks.


# Setup screen.
screen = turtle.Screen()
screen.setup(width=SA_X, height=SA_Y)
turtle.resizemode('user')
screen.title("Search Pattern")
rand_x = random.randint(0, int(SA_X / 2)) * random.choice([-1, 1])
rand_y = random.randint(0, int(SA_Y / 2)) * random.choice([-1, 1])
```

```
# Set up turtle images.
seaman_image = 'seaman.gif'
screen.addshape(seaman_image)
copter_image_left = 'helicopter_left.gif'
copter_image_right = 'helicopter_right.gif'
screen.addshape(copter_image_left)
screen.addshape(copter_image_right)

# Instantiate seaman turtle.
seaman = turtle.Turtle(seaman_image)
seaman.hideturtle()
seaman.penup()
seaman.setpos(rand_x, rand_y)
seaman.showturtle()

# Instantiate copter turtle.
turtle.shape(copter_image_right)
turtle.hideturtle()
turtle.pencolor('black')
turtle.penup()
turtle.setpos(-(int(SA_X / 2) - TRACK_SPACING), int(SA_Y / 2) - TRACK_SPACING)
turtle.showturtle()
turtle.pendown()

# Run search pattern and announce discovery of seaman.
for i in range(int(SA_Y / TRACK_SPACING)):
    turtle.fd(SA_X - TRACK_SPACING * 2)
    turtle.rt(90)
    turtle.fd(TRACK_SPACING / 2)
    turtle.rt(90)
    turtle.shape(copter_image_left)
    turtle.fd(SA_X - TRACK_SPACING * 2)
    turtle.lt(90)
    turtle.fd(TRACK_SPACING / 2)
    turtle.lt(90)
    turtle.shape(copter_image_right)
    if turtle.ycor() - seaman.ycor() <= 10:
        turtle.write("     Seaman found!",
                     align='left',
                     font=("Arial", 15, 'normal', 'bold', 'italic'))
        time.sleep(3)

        break
```

Start Me Up!

*practice_grav
_assist_stationary
.py*

```
"""gravity_assist_stationary.py

Moon approaches stationary ship, which is swung around and flung away.

Credit: Eric T. Mortenson
"""
```

```
from turtle import Shape, Screen, Turtle, Vec2D as Vec
import turtle
import math

# User input:
G = 8  # Gravitational constant used for the simulation.
NUM_LOOPS = 4100  # Number of time steps in simulation.
Ro_X = 0  # Ship starting position x coordinate.
Ro_Y = -50  # Ship starting position y coordinate.
Vo_X = 0  # Ship velocity x component.
Vo_Y = 0  # Ship velocity y component.


MOON_MASS = 1_250_000


class GravSys():
    """Runs a gravity simulation on n-bodies."""

    def __init__(self):
        self.bodies = []
        self.t = 0
        self.dt = 0.001

    def sim_loop(self):
        """Loop bodies in a list through time steps."""
        for _ in range(NUM_LOOPS):
            self.t += self.dt
            for body in self.bodies:
                body.step()


class Body(Turtle):
    """Celestial object that orbits and projects gravity field."""
    def __init__(self, mass, start_loc, vel, gravsys, shape):
        super().__init__(shape=shape)
        self.gravsys = gravsys
        self.penup()
        self.mass=mass
        self.setpos(start_loc)
        self.vel = vel
        gravsys.bodies.append(self)
        self.pendown()  # uncomment to draw path behind object

    def acc(self):
        """Calculate combined force on body and return vector components."""
        a = Vec(0,0)
        for body in self.gravsys.bodies:
            if body != self:
                r = body.pos() - self.pos()
                a += (G * body.mass / abs(r)**3) * r  # units dist/time^2
        return a
```

```python
    def step(self):
        """Calculate position, orientation, and velocity of a body."""
        dt = self.gravsys.dt
        a = self.acc()
        self.vel = self.vel + dt * a
        xOld, yOld = self.pos()  # for orienting ship
        self.setpos(self.pos() + dt * self.vel)
        xNew, yNew = self.pos()  # for orienting ship
        if self.gravsys.bodies.index(self) == 1: # the CSM
            dir_radians = math.atan2(yNew-yOld,xNew-xOld)  # for orienting ship
            dir_degrees = dir_radians * 180 / math.pi # for orienting ship
            self.setheading(dir_degrees+90)  # for orienting ship


def main():
    # Setup screen
    screen = Screen()
    screen.setup(width=1.0, height=1.0)  # for fullscreen
    screen.bgcolor('black')
    screen.title("Gravity Assist Example")

    # Instantiate gravitational system
    gravsys = GravSys()

    # Instantiate Planet
    image_moon = 'moon_27x27.gif'
    screen.register_shape(image_moon)
    moon = Body(MOON_MASS, (500, 0), Vec(-500, 0), gravsys, image_moon)
    moon.pencolor('gray')

    # Build command-service-module (csm) shape
    csm = Shape('compound')
    cm = ((0, 30), (0, -30), (30, 0))
    csm.addcomponent(cm, 'red', 'red')
    sm = ((-60,30), (0, 30), (0, -30), (-60, -30))
    csm.addcomponent(sm, 'red', 'black')
    nozzle = ((-55, 0), (-90, 20), (-90, -20))
    csm.addcomponent(nozzle, 'red', 'red')
    screen.register_shape('csm', csm)

    # Instantiate Apollo 8 CSM turtle
    ship = Body(1, (Ro_X, Ro_Y), Vec(Vo_X, Vo_Y), gravsys, "csm")
    ship.shapesize(0.2)
    ship.color('red') # path color
    ship.getscreen().tracer(1, 0)
    ship.setheading(90)

    gravsys.sim_loop()

if __name__=='__main__':
    main()
```

Shut Me Down!

```
"""gravity_assist_intersecting.py

Moon and ship cross orbits and moon slows and turns ship.

Credit: Eric T. Mortenson
"""
from turtle import Shape, Screen, Turtle, Vec2D as Vec
import turtle
import math
import sys

# User input:
G = 8  # Gravitational constant used for the simulation.
NUM_LOOPS = 7000  # Number of time steps in simulation.
Ro_X = -152.18  # Ship starting position x coordinate.
Ro_Y = 329.87  # Ship starting position y coordinate.
Vo_X = 423.10  # Ship translunar injection velocity x component.
Vo_Y = -512.26  # Ship translunar injection velocity y component.

MOON_MASS = 1_250_000


class GravSys():
    """Runs a gravity simulation on n-bodies."""

    def __init__(self):
        self.bodies = []
        self.t = 0
        self.dt = 0.001

    def sim_loop(self):
        """Loop bodies in a list through time steps."""
        for index in range(NUM_LOOPS): # stops simulation after while
            self.t += self.dt
            for body in self.bodies:
                body.step()


class Body(Turtle):
    """Celestial object that orbits and projects gravity field."""
    def __init__(self, mass, start_loc, vel, gravsys, shape):
        super().__init__(shape=shape)
        self.gravsys = gravsys
        self.penup()
        self.mass=mass
        self.setpos(start_loc)
        self.vel = vel
        gravsys.bodies.append(self)
        self.pendown()  # uncomment to draw path behind object
```

```
def acc(self):
    """Calculate combined force on body and return vector components."""
    a = Vec(0,0)
    for body in self.gravsys.bodies:
        if body != self:
            r = body.pos() - self.pos()
            a += (G * body.mass / abs(r)**3) * r  # units dist/time^2
    return a

def step(self):
    """Calculate position, orientation, and velocity of a body."""
    dt = self.gravsys.dt
    a = self.acc()
    self.vel = self.vel + dt * a
    xOld, yOld = self.pos()  # for orienting ship
    self.setpos(self.pos() + dt * self.vel)
    xNew, yNew = self.pos()  # for orienting ship
    if self.gravsys.bodies.index(self) == 1:  # the CSM
        dir_radians = math.atan2(yNew-yOld,xNew-xOld)  # for orienting ship
        dir_degrees = dir_radians * 180 / math.pi # for orienting ship
        self.setheading(dir_degrees+90)  # for orienting ship


def main():
    # Setup screen
    screen = Screen()
    screen.setup(width=1.0, height=1.0)  # for fullscreen
    screen.bgcolor('black')
    screen.title("Gravity Assist Example")

    # Instantiate gravitational system
    gravsys = GravSys()

    # Instantiate Planet
    image_moon = 'moon_27x27.gif'
    screen.register_shape(image_moon)
    moon = Body(MOON_MASS, (-250, 0), Vec(500, 0), gravsys, image_moon)
    moon.pencolor('gray')

    # Build command-service-module (csm) shape
    csm = Shape('compound')
    cm = ((0, 30), (0, -30), (30, 0))
    csm.addcomponent(cm, 'red', 'red')
    sm = ((-60,30), (0, 30), (0, -30), (-60, -30))
    csm.addcomponent(sm, 'red', 'black')
    nozzle = ((-55, 0), (-90, 20), (-90, -20))
    csm.addcomponent(nozzle, 'red', 'red')
    screen.register_shape('csm', csm)

    # Instantiate Apollo 8 CSM turtle
    ship = Body(1, (Ro_X, Ro_Y), Vec(Vo_X, Vo_Y), gravsys, "csm")
    ship.shapesize(0.2)
    ship.color('red')  # path color
    ship.getscreen().tracer(1, 0)
    ship.setheading(90)
```

```
        gravsys.sim_loop()

if __name__=='__main__':
    main()
```

___

Chapter 7: Selecting Martian Landing Sites

Confirming That Drawings Become Part of an Image

___

practice_confirm
_drawing_part_of
_image.py

```python
"""Test that drawings become part of an image in OpenCV."""
import numpy as np
import cv2 as cv

IMG = cv.imread('mola_1024x501.png', cv.IMREAD_GRAYSCALE)

ul_x, ul_y = 0, 167
lr_x, lr_y = 32, 183
rect_img = IMG[ul_y : lr_y, ul_x : lr_x]

def run_stats(image):
    """Run stats on a numpy array made from an image."""
    print('mean = {}'.format(np.mean(image)))
    print('std = {}'.format(np.std(image)))
    print('ptp = {}'.format(np.ptp(image)))
    print()
    cv.imshow('img', IMG)
    cv.waitKey(1000)

# Stats with no drawing on screen:
print("No drawing")
run_stats(rect_img)

# Stats with white rectangle outline:
print("White outlined rectangle")
cv.rectangle(IMG, (ul_x, ul_y), (lr_x, lr_y), (255, 0, 0), 1)
run_stats(rect_img)

# Stats with rectangle filled with white:
print("White-filled rectangle")
cv.rectangle(IMG, (ul_x, ul_y), (lr_x, lr_y), (255, 0, 0), -1)
run_stats(rect_img)
```

___

Extracting an Elevation Profile

___

practice_profile
_olympus.py

```python
"""West-East elevation profile through Olympus Mons."""
from PIL import Image, ImageDraw
from matplotlib import pyplot as plt
```

```
# Load image and get x and z values along horiz profile parallel to y _coord.
y_coord = 202
im = Image.open('mola_1024x512_200mp.jpg').convert('L')
width, height = im.size
x_vals = [x for x in range(width)]
z_vals = [im.getpixel((x, y_coord)) for x in x_vals]

# Draw profile on MOLA image.
draw = ImageDraw.Draw(im)
draw.line((0, y_coord, width, y_coord), fill=255, width=3)
draw.text((100, 165), 'Olympus Mons', fill=255)
im.show()

# Make profile plot.
fig, ax = plt.subplots(figsize=(9, 4))
axes = plt.gca()
axes.set_ylim(0, 400)
ax.plot(x_vals, z_vals, color='black')
ax.set(xlabel='x-coordinate',
       ylabel='Intensity (height)',
       title="Mars Elevation Profile (y = 202)")
ratio = 0.15   # Reduces vertical exaggeration in profile.
xleft, xright = ax.get_xlim()
ybase, ytop = ax.get_ylim()
ax.set_aspect(abs((xright-xleft)/(ybase-ytop)) * ratio)
plt.text(0, 310, 'WEST', fontsize=10)
plt.text(980, 310, 'EAST', fontsize=10)
plt.text(100, 280, 'Olympus Mons', fontsize=8)
##ax.grid()
plt.show()
```

Plotting in 3D

*practice_3d
_plotting.py*

```
"""Plot Mars MOLA map image in 3D.  Credit Eric T. Mortenson."""
import numpy as np
import cv2 as cv
import matplotlib.pyplot as plt
from mpl_toolkits import mplot3d

IMG_GRAY = cv.imread('mola_1024x512_200mp.jpg', cv.IMREAD_GRAYSCALE)

x = np.linspace(1023, 0, 1024)
y = np.linspace(0, 511, 512)

X, Y = np.meshgrid(x, y)
Z = IMG_GRAY[0:512, 0:1024]

fig = plt.figure()
ax = plt.axes(projection='3d')
ax.contour3D(X, Y, Z, 150, cmap='gist_earth')  # 150=number of contours
ax.auto_scale_xyz([1023, 0], [0, 511], [0, 500])
plt.show()
```

Mixing Maps

This practice project uses two programs, *practice_geo_map_step_1of2.py* and *practice_geo_map_step_2of2.py*, that must be run in order.

practice_geo_map_step_1of2.py

*practice_geo_map_
_step_1of2.py*

```
"""Threshold a grayscale image using pixel values and save to file."""
import cv2 as cv

IMG_GEO = cv.imread('Mars_Global_Geology_Mariner9_1024.jpg',
                    cv.IMREAD_GRAYSCALE)
cv.imshow('map', IMG_GEO)
cv.waitKey(1000)
img_copy = IMG_GEO.copy()
lower_limit = 170  # Lowest grayscale value for volcanic deposits
upper_limit = 185  # Highest grayscale value for volcanic deposits

# Using 1024 x 512 image
for x in range(1024):
    for y in range(512):
        if lower_limit <= img_copy[y, x] <= upper_limit:
            img_copy[y, x] = 1  # Set to 255 to visualize results.
        else:
            img_copy[y, x] = 0

cv.imwrite('geo_thresh.jpg', img_copy)
cv.imshow('thresh', img_copy)
cv.waitKey(0)
```

practice_geo_map_step_2of2.py

*practice_geo_map_
_step_2of2.py*

```
"""Select Martian landing sites based on surface smoothness and geology."""
import tkinter as tk
from PIL import Image, ImageTk
import numpy as np
import cv2 as cv

# CONSTANTS: User Input:
IMG_GRAY = cv.imread('mola_1024x512_200mp.jpg', cv.IMREAD_GRAYSCALE)
IMG_GEO = cv.imread('geo_thresh.jpg', cv.IMREAD_GRAYSCALE)
IMG_COLOR = cv.imread('mola_color_1024x506.png')
RECT_WIDTH_KM = 670  # Site rectangle width in kilometers.
RECT_HT_KM = 335  # Site rectangle height in kilometers.
MIN_ELEV_LIMIT = 60  # Intensity values (0-255).
MAX_ELEV_LIMIT = 255
NUM_CANDIDATES = 20  # Number of candidate landing sites to display.

#--------------------------------------------------------------------------
```

```python
# CONSTANTS: Derived and fixed:
IMG_GRAY_GEO = IMG_GRAY * IMG_GEO
IMG_HT, IMG_WIDTH = IMG_GRAY.shape
MARS_CIRCUM = 21344  # Circumference in kilometers.
PIXELS_PER_KM = IMG_WIDTH / MARS_CIRCUM
RECT_WIDTH = int(PIXELS_PER_KM * RECT_WIDTH_KM)
RECT_HT = int(PIXELS_PER_KM * RECT_HT_KM)
LAT_30_N = int(IMG_HT / 3)
LAT_30_S = LAT_30_N * 2
STEP_X = int(RECT_WIDTH / 2)  # Dividing by 4 yields more rect choices
STEP_Y = int(RECT_HT / 2)  # Dividing by 4 yields more rect choices

# Create tkinter screen and drawing canvas
screen = tk.Tk()
canvas = tk.Canvas(screen, width=IMG_WIDTH, height=IMG_HT + 130)


class Search():
    """Read image and identify landing sites based on input criteria."""

    def __init__(self, name):
        self.name = name
        self.rect_coords = {}
        self.rect_means = {}
        self.rect_ptps = {}
        self.rect_stds = {}
        self.ptp_filtered = []
        self.std_filtered = []
        self.high_graded_rects = []

    def run_rect_stats(self):
        """Define rectangular search areas and calculate internal stats."""
        ul_x, ul_y = 0, LAT_30_N
        lr_x, lr_y = RECT_WIDTH, LAT_30_N + RECT_HT
        rect_num = 1

        while True:
            rect_img = IMG_GRAY_GEO[ul_y : lr_y, ul_x : lr_x]
            self.rect_coords[rect_num] = [ul_x, ul_y, lr_x, lr_y]
            if MAX_ELEV_LIMIT >= np.mean(rect_img) >= MIN_ELEV_LIMIT:
                self.rect_means[rect_num] = np.mean(rect_img)
                self.rect_ptps[rect_num] = np.ptp(rect_img)
                self.rect_stds[rect_num] = np.std(rect_img)
            rect_num += 1

            # Move the rectangle.
            ul_x += STEP_X
            lr_x = ul_x + RECT_WIDTH
            if lr_x > IMG_WIDTH:
                ul_x = 0
```

```
                ul_y += STEP_Y
                lr_x = RECT_WIDTH
                lr_y += STEP_Y
            if lr_y > LAT_30_S + STEP_Y:
                break

    def draw_qc_rects(self):
        """Draw overlapping search rectangles on image as a check."""
        img_copy = IMG_GRAY_GEO.copy()
        rects_sorted = sorted(self.rect_coords.items(), key=lambda x: x[0])
        print("\nRect Number and Corner Coordinates (ul_x, ul_y, lr_x, lr_y):")
        for k, v in rects_sorted:
            print("rect: {}, coords: {}".format(k, v))
            cv.rectangle(img_copy,
                        (self.rect_coords[k][0], self.rect_coords[k][1]),
                        (self.rect_coords[k][2], self.rect_coords[k][3]),
                        (255, 0, 0), 1)
        cv.imshow('QC Rects {}'.format(self.name), img_copy)
        cv.waitKey(3000)
        cv.destroyAllWindows()

    def sort_stats(self):
        """Sort dictionaries by values and create lists of top N keys."""
        ptp_sorted = (sorted(self.rect_ptps.items(), key=lambda x: x[1]))
        self.ptp_filtered = [x[0] for x in ptp_sorted[:NUM_CANDIDATES]]
        std_sorted = (sorted(self.rect_stds.items(), key=lambda x: x[1]))
        self.std_filtered = [x[0] for x in std_sorted[:NUM_CANDIDATES]]

        # Make list of rects where filtered std & ptp coincide.
        for rect in self.std_filtered:
            if rect in self.ptp_filtered:
                self.high_graded_rects.append(rect)

    def draw_filtered_rects(self, image, filtered_rect_list):
        """Draw rectangles in list on image and return image."""
        img_copy = image.copy()
        for k in filtered_rect_list:
            cv.rectangle(img_copy,
                        (self.rect_coords[k][0], self.rect_coords[k][1]),
                        (self.rect_coords[k][2], self.rect_coords[k][3]),
                        (255, 0, 0), 1)
            cv.putText(img_copy, str(k),
                        (self.rect_coords[k][0] + 1, self.rect_coords[k][3]- 1),
                        cv.FONT_HERSHEY_PLAIN, 0.65, (255, 0, 0), 1)

        # Draw latitude limits.
        cv.putText(img_copy, '30 N', (10, LAT_30_N - 7),
                    cv.FONT_HERSHEY_PLAIN, 1, 255)
        cv.line(img_copy, (0, LAT_30_N), (IMG_WIDTH, LAT_30_N),
                (255, 0, 0), 1)
        cv.line(img_copy, (0, LAT_30_S), (IMG_WIDTH, LAT_30_S),
                (255, 0, 0), 1)
```

```
            cv.putText(img_copy, '30 S', (10, LAT_30_S + 16),
                       cv.FONT_HERSHEY_PLAIN, 1, 255)

        return img_copy

    def make_final_display(self):
        """Use Tk to show map of final rects & printout of their statistics."""
        screen.title('Sites by MOLA Gray STD & PTP {} Rect'.format(self.name))
        # Draw the high-graded rects on the colored elevation map.
        img_color_rects = self.draw_filtered_rects(IMG_COLOR,
                                                   self.high_graded_rects)
        # Convert image from CV BGR to RGB for use with Tkinter.
        img_converted = cv.cvtColor(img_color_rects, cv.COLOR_BGR2RGB)
        img_converted = ImageTk.PhotoImage(Image.fromarray(img_converted))
        canvas.create_image(0, 0, image=img_converted, anchor=tk.NW)
        # Add stats for each rectangle at bottom of canvas.
        txt_x = 5
        txt_y = IMG_HT + 15
        for k in self.high_graded_rects:
            canvas.create_text(txt_x, txt_y, anchor='w', font=None,
                               text=
                               "rect={}  mean elev={:.1f}  std={:.2f}  ptp={}"
                               .format(k, self.rect_means[k],
                                       self.rect_stds[k],
                                       self.rect_ptps[k]))
            txt_y += 15
            if txt_y >= int(canvas.cget('height')) - 10:
                txt_x += 300
                txt_y = IMG_HT + 15
        canvas.pack()
        screen.mainloop()

def main():
    app = Search('670x335 km')
    app.run_rect_stats()
    app.draw_qc_rects()
    app.sort_stats()
    ptp_img = app.draw_filtered_rects(IMG_GRAY_GEO, app.ptp_filtered)
    std_img = app.draw_filtered_rects(IMG_GRAY_GEO, app.std_filtered)

    # Display filtered rects on grayscale map.
    cv.imshow('Sorted by ptp for {} rect'.format(app.name), ptp_img)
    cv.waitKey(3000)
    cv.imshow('Sorted by std for {} rect'.format(app.name), std_img)
    cv.waitKey(3000)

    app.make_final_display()  # includes call to mainloop()

if __name__ == '__main__':
    main()
```

Chapter 8: Detecting Distant Exoplanets

Detecting Alien Megastructures

```python
"""Simulate transit of alien array and plot light curve."""
import numpy as np
import cv2 as cv
import matplotlib.pyplot as plt

IMG_HT = 400
IMG_WIDTH = 500
BLACK_IMG = np.zeros((IMG_HT, IMG_WIDTH), dtype='uint8')
STAR_RADIUS = 165
EXO_START_X = -250
EXO_START_Y = 150
EXO_DX = 3
NUM_FRAMES = 500

def main():
    intensity_samples = record_transit(EXO_START_X, EXO_START_Y)
    rel_brightness = calc_rel_brightness(intensity_samples)
    plot_light_curve(rel_brightness)

def record_transit(exo_x, exo_y):
    """Draw array transiting star and return list of intensity changes."""
    intensity_samples = []
    for _ in range(NUM_FRAMES):
        temp_img = BLACK_IMG.copy()
        # Draw star:
        cv.circle(temp_img, (int(IMG_WIDTH / 2), int(IMG_HT / 2)),
                  STAR_RADIUS, 255, -1)
        # Draw alien array:
        cv.rectangle(temp_img, (exo_x, exo_y),
                     (exo_x + 20, exo_y + 140), 0, -1)
        cv.rectangle(temp_img, (exo_x - 360, exo_y),
                     (exo_x + 10, exo_y + 140), 0, 5)
        cv.rectangle(temp_img, (exo_x - 380, exo_y),
                     (exo_x - 310, exo_y + 140), 0, -1)
        intensity = temp_img.mean()
        cv.putText(temp_img, 'Mean Intensity = {}'.format(intensity), (5, 390),
                   cv.FONT_HERSHEY_PLAIN, 1, 255)
        cv.imshow('Transit', temp_img)
        cv.waitKey(10)
        intensity_samples.append(intensity)
        exo_x += EXO_DX
    return intensity_samples

def calc_rel_brightness(intensity_samples):
    """Return list of relative brightness from list of intensity values."""
    rel_brightness = []
    max_brightness = max(intensity_samples)
    for intensity in intensity_samples:
        rel_brightness.append(intensity / max_brightness)
```

```
        return rel_brightness

    def plot_light_curve(rel_brightness):
        """Plot changes in relative brightness vs. time."""
        plt.plot(rel_brightness, color='red', linestyle='dashed',
                 linewidth=2)
        plt.title('Relative Brightness vs. Time')
        plt.xlim(-150, 500)
        plt.show()

    if __name__ == '__main__':
        main()
```

Detecting Asteroid Transits

practice
_asteroids.py

```
"""Simulate transit of asteroids and plot light curve."""
import random
import numpy as np
import cv2 as cv
import matplotlib.pyplot as plt

STAR_RADIUS = 165
BLACK_IMG = np.zeros((400, 500, 1), dtype="uint8")
NUM_ASTEROIDS = 15
NUM_LOOPS = 170


class Asteroid():
    """Draws a circle on an image that represents an asteroid."""

    def __init__(self, number):
        self.radius = random.choice((1, 1, 1, 1, 1, 1, 1, 1, 1, 1, 2, 2, 2, 3))
        self.x = random.randint(-30, 60)
        self.y = random.randint(220, 230)
        self.dx = 3

    def move_asteroid(self, image):
        """Draw and move asteroid object."""
        cv.circle(image, (self.x, self.y), self.radius, 0, -1)
        self.x += self.dx


def record_transit(start_image):
    """Simulate transit of asteroids over star and return intensity list."""
    asteroid_list = []
    intensity_samples = []

    for i in range(NUM_ASTEROIDS):
        asteroid_list.append(Asteroid(i))

    for _ in range(NUM_LOOPS):
        temp_img = start_image.copy()
        # Draw star.
```

```
            cv.circle(temp_img, (250, 200), STAR_RADIUS, 255, -1)
            for ast in asteroid_list:
                ast.move_asteroid(temp_img)
            intensity = temp_img.mean()
            cv.putText(temp_img, 'Mean Intensity = {}'.format(intensity),
                    (5, 390), cv.FONT_HERSHEY_PLAIN, 1, 255)
            cv.imshow('Transit', temp_img)
            intensity_samples.append(intensity)
            cv.waitKey(50)
        cv.destroyAllWindows()
        return intensity_samples

    def calc_rel_brightness(image):
        """Calculate and return list of relative brightness samples."""
        rel_brightness = record_transit(image)
        max_brightness = max(rel_brightness)
        for i, j in enumerate(rel_brightness):
            rel_brightness[i] = j / max_brightness
        return rel_brightness

    def plot_light_curve(rel_brightness):
        "Plot light curve from relative brightness list."""
        plt.plot(rel_brightness, color='red', linestyle='dashed',
                linewidth=2, label='Relative Brightness')
        plt.legend(loc='upper center')
        plt.title('Relative Brightness vs. Time')
        plt.show()

    relative_brightness = calc_rel_brightness(BLACK_IMG)
    plot_light_curve(relative_brightness)
```

Incorporating Limb Darkening

practice_limb
_darkening.py

```
    """Simulate transit of exoplanet, plot light curve, estimate planet radius."""
    import cv2 as cv
    import matplotlib.pyplot as plt

    IMG_HT = 400
    IMG_WIDTH = 500
    BLACK_IMG = cv.imread('limb_darkening.png', cv.IMREAD_GRAYSCALE)
    EXO_RADIUS = 7
    EXO_START_X = 40
    EXO_START_Y = 230
    EXO_DX = 3
    NUM_FRAMES = 145

    def main():
        intensity_samples = record_transit(EXO_START_X, EXO_START_Y)
        relative_brightness = calc_rel_brightness(intensity_samples)
        plot_light_curve(relative_brightness)

    def record_transit(exo_x, exo_y):
        """Draw planet transiting star and return list of intensity changes."""
```

```
            intensity_samples = []
            for _ in range(NUM_FRAMES):
                temp_img = BLACK_IMG.copy()
                # Draw exoplanet:
                cv.circle(temp_img, (exo_x, exo_y), EXO_RADIUS, 0, -1)
                intensity = temp_img.mean()
                cv.putText(temp_img, 'Mean Intensity = {}'.format(intensity), (5, 390),
                           cv.FONT_HERSHEY_PLAIN, 1, 255)
                cv.imshow('Transit', temp_img)
                cv.waitKey(30)
                intensity_samples.append(intensity)
                exo_x += EXO_DX
            return intensity_samples

        def calc_rel_brightness(intensity_samples):
            """Return list of relative brightness from list of intensity values."""
            rel_brightness = []
            max_brightness = max(intensity_samples)
            for intensity in intensity_samples:
                rel_brightness.append(intensity / max_brightness)
            return rel_brightness

        def plot_light_curve(rel_brightness):
            """Plot changes in relative brightness vs. time."""
            plt.plot(rel_brightness, color='red', linestyle='dashed',
                     linewidth=2, label='Relative Brightness')
            plt.legend(loc='upper center')
            plt.title('Relative Brightness vs. Time')
        ##    plt.ylim(0.995, 1.001)
            plt.show()

        if __name__ == '__main__':
            main()
```

Detecting an Alien Armada

*practice_alien
_armada.py*

```
"""Simulate transit of alien armada with light curve."""
import random
import numpy as np
import cv2 as cv
import matplotlib.pyplot as plt

STAR_RADIUS = 165
BLACK_IMG = np.zeros((400, 500, 1), dtype="uint8")
NUM_SHIPS = 5
NUM_LOOPS = 300  # Number of simulation frames to run


class Ship():
    """Draws and moves a ship object on an image."""

    def __init__(self, number):
        self.number = number
```

```
                self.shape = random.choice(['>>>|==H[X]',
                                            '>>|==H[XX}=))-',
                                            '>>|==H[XX]=(-'])
                self.size = random.choice([0.7, 0.8, 1])
                self.x = random.randint(-180, -80)
                self.y = random.randint(80, 350)
                self.dx = random.randint(2, 4)

        def move_ship(self, image):
            """Draws and moves ship object."""
            font = cv.FONT_HERSHEY_PLAIN
            cv.putText(img=image,
                        text=self.shape,
                        org=(self.x, self.y),
                        fontFace=font,
                        fontScale=self.size,
                        color=0,
                        thickness=5)
            self.x += self.dx


    def record_transit(start_image):
        """Runs simulation and returns list of intensity measurements per frame."""
        ship_list = []
        intensity_samples = []

        for i in range(NUM_SHIPS):
            ship_list.append(Ship(i))

        for _ in range(NUM_LOOPS):
            temp_img = start_image.copy()
            cv.circle(temp_img, (250, 200), STAR_RADIUS, 255, -1)  # The star.
            for ship in ship_list:
                ship.move_ship(temp_img)
            intensity = temp_img.mean()
            cv.putText(temp_img, 'Mean Intensity = {}'.format(intensity),
                        (5, 390), cv.FONT_HERSHEY_PLAIN, 1, 255)
            cv.imshow('Transit', temp_img)
            intensity_samples.append(intensity)
            cv.waitKey(50)
        cv.destroyAllWindows()
        return intensity_samples

    def calc_rel_brightness(image):
        """Return list of relative brightness measurments for planetary transit."""
        rel_brightness = record_transit(image)
        max_brightness = max(rel_brightness)
        for i, j in enumerate(rel_brightness):
            rel_brightness[i] = j / max_brightness
        return rel_brightness

    def plot_light_curve(rel_brightness):
        """Plots curve of relative brightness vs. time."""
        plt.plot(rel_brightness, color='red', linestyle='dashed',
                linewidth=2, label='Relative Brightness')
```

```
        plt.legend(loc='upper center')
        plt.title('Relative Brightness vs. Time')
        plt.show()

    relative_brightness = calc_rel_brightness(BLACK_IMG)
    plot_light_curve(relative_brightness)
```

Detecting a Planet with a Moon

practice_planet
_moon.py

```python
"""Moon animation credit Eric T. Mortenson."""
import math
import numpy as np
import cv2 as cv
import matplotlib.pyplot as plt


IMG_HT = 500
IMG_WIDTH = 500
BLACK_IMG = np.zeros((IMG_HT, IMG_WIDTH, 1), dtype='uint8')
STAR_RADIUS = 200
EXO_RADIUS = 20
EXO_START_X = 20
EXO_START_Y = 250
MOON_RADIUS = 5
NUM_DAYS = 200   # number days in year

def main():
    intensity_samples = record_transit(EXO_START_X, EXO_START_Y)
    relative_brightness = calc_rel_brightness(intensity_samples)
    print('\nestimated exoplanet radius = {:.2f}\n'
          .format(STAR_RADIUS * math.sqrt(max(relative_brightness)
                                          -min(relative_brightness))))
    plot_light_curve(relative_brightness)

def record_transit(exo_x, exo_y):
    """Draw planet transiting star and return list of intensity changes."""
    intensity_samples = []
    for dt in range(NUM_DAYS):
        temp_img = BLACK_IMG.copy()
        # Draw star:
        cv.circle(temp_img, (int(IMG_WIDTH / 2), int(IMG_HT/2)),
                  STAR_RADIUS, 255, -1)
        # Draw exoplanet
        cv.circle(temp_img, (int(exo_x), int(exo_y)), EXO_RADIUS, 0, -1)
        # Draw moon
        if dt != 0:
            cv.circle(temp_img, (int(moon_x), int(moon_y)), MOON_RADIUS, 0, -1)
        intensity = temp_img.mean()
        cv.putText(temp_img, 'Mean Intensity = {}'.format(intensity), (5, 10),
                   cv.FONT_HERSHEY_PLAIN, 1, 255)
        cv.imshow('Transit', temp_img)
        cv.waitKey(10)
        intensity_samples.append(intensity)
        exo_x = IMG_WIDTH / 2 - (IMG_WIDTH / 2 - 20) * \
```

```
                    math.cos(2 * math.pi * dt / (NUM_DAYS)*(1 / 2))
        moon_x = exo_x + \
                  3 * EXO_RADIUS * math.sin(2 * math.pi * dt / NUM_DAYS *(5))
        moon_y = IMG_HT / 2 - \
                  0.25 * EXO_RADIUS * \
                  math.sin(2 * math.pi * dt / NUM_DAYS * (5))
    cv.destroyAllWindows()

    return intensity_samples

def calc_rel_brightness(intensity_samples):
    """Return list of relative brightness from list of intensity values."""
    rel_brightness = []
    max_brightness = max(intensity_samples)
    for intensity in intensity_samples:
        rel_brightness.append(intensity / max_brightness)
    return rel_brightness

def plot_light_curve(rel_brightness):
    """Plot changes in relative brightness vs. time."""
    plt.plot(rel_brightness, color='red', linestyle='dashed',
             linewidth=2, label='Relative Brightness')
    plt.legend(loc='upper center')
    plt.title('Relative Brightness vs. Time')
    plt.show()

if __name__ == '__main__':
    main()
```

Figure A-1 summarizes the output from the *practice_planet_moon.py* program.

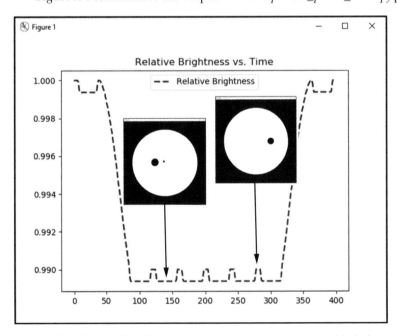

Figure A-1: Light curve for planet and moon where moon passes behind planet

Measuring the Length of an Exoplanet's Day

practice_length _of_day.py

```
"""Read-in images, calculate mean intensity, plot relative intensity vs time."""
import os
from statistics import mean
import cv2 as cv
import numpy as np
import matplotlib.pyplot as plt
from scipy import signal  # See Chap. 1 to install scipy.

# Switch to the folder containing images.
os.chdir('br549_pixelated')
images = sorted(os.listdir())
intensity_samples = []

# Convert images to grayscale and make a list of mean intensity values.
for image in images:
    img = cv.imread(image, cv.IMREAD_GRAYSCALE)
    intensity = img.mean()
    intensity_samples.append(intensity)

# Generate a list of relative intensity values.
rel_intensity = intensity_samples[:]
max_intensity = max(rel_intensity)
for i, j in enumerate(rel_intensity):
    rel_intensity[i] = j / max_intensity

# Plot relative intensity values vs frame number (time proxy).
plt.plot(rel_intensity, color='red', marker='o', linestyle='solid',
         linewidth=2, markersize=0, label='Relative Intensity')
plt.legend(loc='upper center')
plt.title('Exoplanet BR549 Relative Intensity vs. Time')
plt.ylim(0.8, 1.1)
plt.xticks(np.arange(0, 50, 5))
plt.grid()
print("\nManually close plot window after examining to continue program.")
plt.show()

# Find period / length of day.
# Estimate peak height and separation (distance) limits from plot.
# height and distance parameters represent >= limits.
peaks = signal.find_peaks(rel_intensity, height=0.95, distance=5)
print(f"peaks = {peaks}")
print("Period = {}".format(mean(np.diff(peaks[0]))))
```

Chapter 9: Identifying Friend or Foe

Blurring Faces

practice_blur.py

```
import cv2 as cv

path = "C:/Python372/Lib/site-packages/cv2/data/"
face_cascade = cv.CascadeClassifier(path + 'haarcascade_frontalface_alt.xml')

cap = cv.VideoCapture(0)

while True:
    _, frame = cap.read()
    face_rects = face_cascade.detectMultiScale(frame, scaleFactor=1.2,
                                               minNeighbors=3)

    for (x, y, w, h) in face_rects:
        face = cv.blur(frame[y:y + h, x:x + w], (25, 25))
        frame[y:y + h, x: x + w] = face
        cv.rectangle(frame, (x,y), (x+w, y+h), (0, 255, 0), 2)

    cv.imshow('frame', frame)
    if cv.waitKey(1) & 0xFF == ord('q'):
        break

cap.release()
cv.destroyAllWindows()
```

Chapter 10: Restricting Access with Face Recognition

Challenge Project: Adding a Password and Video Capture

The following snippet addresses the part of the challenge project concerned with recognizing faces from a video stream.

*challenge_video
_recognize.py*

```
"""Recognize Capt. Demming's face in video frame."""
import cv2 as cv

names = {1: "Demming"}

# Set up path to OpenCV's Haar Cascades
path = "C:/Python372/Lib/site-packages/cv2/data/"
detector = cv.CascadeClassifier(path + 'haarcascade_frontalface_default.xml')

# Set up face recognizer and load trained data.
recognizer = cv.face.LBPHFaceRecognizer_create()
recognizer.read('lbph_trainer.yml')
```

```python
# Prepare webcam.
cap = cv.VideoCapture(0)
if not cap.isOpened():
    print("Could not open video device.")
##cap.set(3, 320)  # Frame width.
##cap.set(4, 240)  # Frame height.

while True:
    _, frame = cap.read()
    gray = cv.cvtColor(frame, cv.COLOR_BGR2GRAY)
    face_rects = detector.detectMultiScale(gray,
                                           scaleFactor=1.2,
                                           minNeighbors=5)

    for (x, y, w, h) in face_rects:
        # Resize input so it's closer to training image size.
        gray_resize = cv.resize(gray[y:y + h, x:x + w],
                          (100, 100),
                          cv.INTER_LINEAR)
        predicted_id, dist = recognizer.predict(gray_resize)
        if predicted_id == 1 and dist <= 110:
            name = names[predicted_id]
        else:
            name = 'unknown'
        cv.rectangle(frame, (x, y), (x + w, y + h), (255, 255, 0), 2)
        cv.putText(frame, name, (x + 1, y + h -5),
                   cv.FONT_HERSHEY_SIMPLEX, 0.5, (255, 255, 0), 1)
        cv.imshow('frame', frame)

    if cv.waitKey(1) & 0xFF == ord('q'):
        break

cap.release()
cv.destroyAllWindows()
```

INDEX

C

S

Savage, Martin J., 278
save() method, 261
say() method, 211
scale pyramids, 212
scaled gravitational constants, 133
SciPy package, 7–8
score_sentences() function, 57–59
scraping the web, 53, 62
screen, setting up, 139
Screen subclass, 133
screen updates, 144
search
    calculating effectiveness, 16–17
    conducting, 16–17
Search and Rescue project, 5–24
search classes, 10–12
    defining, 10–12, 161
    initializing, 161
search effectiveness probability (SEP), 4
search engine optimization (SEO), 65
Seeing Through a Dog's Eyes project, 281
Selecting Martian Landing Sites
        project, 153–171
select() method, 53
    limits, 55
self.area_actual attribute, 16
self attributes, 12
self parameter
    function, 134
    using, 136
Sending Secrets the WWII Way project,
        93–94, 286–289
sentry guns
    automated, 206
    use of video feeds, 207
sent_tokenize() method, 63
series, 248–249
setpos() method, 135, 272
setup() method, 271
Shape class, 133
shape()function, 109
shapes, 142–143
    building, 142–143
shift value, 84
    overview, 84
Shut Me Down! project, 148, 296–298
SIDE constants, 272
sim_loop() method, 144
Simplifying a Celebration Slideshow
        project, 281

Simulating a Search Pattern project,
        146–147, 292–293
Simulating an Exoplanet Transit
        project, 179–188
simulation hypothesis, 269
simulation loops, 144
sliding window approach, 205
slingshot maneuver, 145
    simulation, 145
Smarter Searches project, 24
Smith, David, 153
sorted() function, 101
sound
    audio recordings, 209, 217
    files, 209
    playsound module, 217
split() function, 36, 89, 191
    tokens, 36
standard deviation
    applying, 154–155
    formula, 154
    sorting, 165
starspots, 200
Start Me Up! project, 147–148, 293–295
statistics.mean() function, 277
statistics module, 276
stemming, 47
step() method, defining, 137
stop words, 39–40, 57–58, 67
    analyzed by natural language
        processing (NLP), 28
    comparing, 39–40
    examples, 54
    functional, 54
    importing, 66
    removing, 57–58
Stopwords Corpus, 30–31
string.replace() method, 56, 89
strings
    f-string format, 211
    Hamming distance, and string
        length, 105
    join() method, turning elements
        into a string with, 55
    length, 105
    ord() function, 220
    string.replace() method, 56, 89
    string.split() method, 89
    text_to_string() function, 33, 34
Structured Query Language, 255

type() function, 67
Tyson, Neil DeGrasse, 269

U

ultra-high energy cosmic rays
(UHECRs), 278
unconstrained faces, 221
underscore (_), 102
Unicode Transformational Format
(UTF), 35, 86
UnicodeDecodeError, 35
while loading text, 67
unit vectors, formula, 136
unknown, 44, 47
unstructured data, 51
unused variables, 102
Pylint, 102
uppercase and lowercase letters,
handling with natural
language processing
(NLP), 58
UTF (Unicode Transformational
Format), 35

V

values
max() function, 185
maximum values, 185
variables, 17, 33, 102, 115
assigning, local, 17, 162
built-in, 22
chi-squared random variable (X 2),
43–44
diff_image variable, 114
excessive, 142
global, 12
__name__ variable, 22
naming, 68
page variable, 54–55
Pylint, 102
unused, 102
Vec2D, 133, 136
vectors, ORB, 104
video feeds
capturing, 232–236
streams, 219–221
virtual environments, xxv
using, xxv
Visual Studio Code, xxii

Visualizing Population Density with a
Choropleth Map project,
246–265
vocab_test() function, defining, 43
vocabularies, analyzed by natural
language processing (NLP),
43–45
voices
changing, 209
Windows OS
American "David," 233
default, 233
default voice, 209
female, 209
male, 209
Vonnegut, Kurt, 48

W

waitKey() method, 14, 107, 184, 235
War of the Worlds, The, 32
warpPerspective() method, 109
webbrowser module, 255, 261
web scraping, 53, 62
Wells, H. G., 28
What a Tangled Web We Weave
project, 281
What's the Difference? project, 120,
290–292
while loops, 19–20, 56, 219
Windows OS, xxii
character encoding, 35
CP-1252, 35
Haar features, 209, 234
holoviews module, 250
PyScripter, xxii
pyttsx3 module, 208
tkinter module, 156
using PowerShell, 42
voices on, 209, 233
word clouds, 64–71
displaying, 70
fine-tuning, 70–71
generating, 67–68
plotting, 69–70
word length
analyzed by natural language
processing (NLP), 28
comparing, 37–39
word_tokenize() method, 30, 35, 57
Wrapping Rectangles project, 175–176
writers, 255

Y

YAML (.yml) files
    overview, 227
    loading, 239
yield statements, 47
    suspending fuctions with, 47

Z

zip() function, 259
Zuber, Maria, 153

Real-World Python is set in New Baskerville, Futura, Dogma, and TheSansMono Condensed. The book was printed and bound by Sheridan Books, Inc. in Chelsea, Michigan. The paper is 60# Finch Offset, which is certified by the Forest Stewardship Council (FSC).

The book uses a layflat binding, in which the pages are bound together with a cold-set, flexible glue and the first and last pages of the resulting book block are attached to the cover. The cover is not actually glued to the book's spine, and when open, the book lies flat and the spine doesn't crack.

RESOURCES

Visit *https://nostarch.com/real-world-python/* for errata and more information.

More no-nonsense books from **NO STARCH PRESS**

IMPRACTICAL PYTHON PROJECTS
Playful Programming Activities to Make You Smarter
by LEE VAUGHAN
424 PP., $29.95
ISBN 978-1-59327-890-8

PYTHON CRASH COURSE, 2ND EDITION
A Hands-On, Project-Based Introduction to Programming
by ERIC MATTHES
544 PP., $39.95
ISBN 978-1-59327-928-8

SERIOUS PYTHON
Black-Belt Advice on Deployment, Scalability, Testing, and More
by JULIEN DANJOU
240 PP., $34.95
ISBN 978-1-59327-878-6

PYTHON BEYOND THE BASICS
Best Practices for Writing Clean Code
by AL SWEIGART
FALL 2020, 286 PP., $34.95
ISBN 978-1-59327-966-0

PYTHON ONE-LINERS
Write Concise, Eloquent Python Like a Professional
by CHRISTIAN MAYER
216 PP., 39.95
ISBN 978-1-7185-0050-1

NATURAL LANGUAGE PROCESSING WITH PYTHON AND SPACY
A Practical Introduction
by YULI VASILIEV
216 PP., $44.95
ISBN 978-1-7185-0052-5

PHONE:
800.420.7240 OR
415.863.9900

EMAIL:
SALES@NOSTARCH.COM

WEB:
WWW.NOSTARCH.COM